The Minerals, Metals & Materials Series

Yuri Hovanski · Yutaka Sato · Piyush Upadhyay ·
Anton A. Naumov · Nilesh Kumar
Editors

Friction Stir Welding and Processing XII

TMS

Springer

Editors
Yuri Hovanski
Brigham Young University
Provo, UT, USA

Yutaka Sato
Tohoku University
Sendai, Japan

Piyush Upadhyay
Pacific Northwest National Laboratory
Richland, WA, USA

Anton A. Naumov
Peter the Great St. Petersburg
Polytechnic University
Saint Petersburg, Russia

Nilesh Kumar
The University of Alabama
Tuscaloosa, AL, USA

ISSN 2367-1181 ISSN 2367-1696 (electronic)
The Minerals, Metals & Materials Series
ISBN 978-3-031-22660-1 ISBN 978-3-031-22661-8 (eBook)
https://doi.org/10.1007/978-3-031-22661-8

Cover Illustration: Left: From Chapter "2D Axisymmetric Modeling of RFSSW Repair and Experimental Validation" Evan Berger et al., Figure 4: Aborted weld, with ferroaluminum compound attached to the tooling. https://doi.org/10.1007/978-3-031-22661-8_12. Right: From Chapter "Submerged Bobbin Tool (SBT) Tunneling Technology" Dwight A. Burford et al., Figure 9: Surface model built from XRM test data results for a channel produced at 635 mm/min (25 ipm) and 750 rpm. https://doi.org/10.1007/978-3-031-22661-8_23. Bottom: From Chapter "The Performance of a Force-Based General Defect Detection Method Outside of Calibration" Johnathon B. Hunt et al., Figure 3: Clamping method and welding table set up for all welds. https://doi.org/10.1007/978-3-031-22661-8_17

This Springer imprint is published by the registered company Springer Nature Switzerland AG
The registered company address is: Gewerbestrasse 11, 6330 Cham, Switzerland

Preface

These proceedings represent the twelfth symposium on Friction Stir Welding and Processing (FSWP) held under the auspices of TMS. This historic proceeding volume represents 32 years of study, research, and implementation since the initial FSW patent was filed in 1991. The continued interest and participation in this symposium and the associated proceedings are an indirect testimony to the growth of this field.

For 2023, a total of 62 abstracts were accepted which includes 8 oral sessions. There are 23 papers included in this volume, which when combined with the previous ten proceedings publications represent more than 350 papers over a 24-year period. These submissions cover all aspects of friction stir technologies including: FSW of high melting temperature materials, FSW of lightweight materials, FSW of dissimilar materials, simulation of FSWP, controls and inspection of FSWP, and derivative technologies like friction stir processing, friction stir spot welding, additive friction stir, and friction stir extrusion. The rise of additive friction stir deposition has shown significant growth over the past few years and, for the first time in this symposium's history, maintains an entire oral session focused on this prolific additive technology.

Friction stir welding was invented by TWI (formerly The Welding Institute), Cambridge, UK, and patented in 1991, although the real growth in this field started several years later. In the last 32 years, FSW has seen significant growth in both technology implementation and scientific exploration. The original patent has led to numerous additional patents issued globally, as various solid-state processing techniques were derived from the original FSW concept. In addition to the tremendous number of derivative technologies that have been developed based on the concept of friction stirring, thousands of papers have been published characterizing and documenting the commercial and scientific benefits of the same.

The organizers would like to thank the Shaping and Forming Committee of the Materials Processing and Manufacturing Division for sponsoring this symposium.

Yuri Hovanski
Yutaka Sato
Piyush Upadhyay
Anton A. Naumov
Nilesh Kumar

Contents

About the Editors

Yuri Hovanski is an associate professor of Manufacturing Engineering at Brigham Young University. He earned a B.S. degree in Mechanical Engineering at Brigham Young University, and then completed his masters and doctorate degrees at Washington State University. As a member of numerous professional societies, he actively participates in ASM, SAE, SME, and TMS serving in numerous leadership roles at the technical committee and division levels. He is a past chair of the TMS Shaping and Forming Committee, the vice-chair of the ASM Joining of Advanced Specialty Materials Committee, and director of the Center of Friction Stir Processing, a retired NSF-IUCRC. He participated in research related to friction stir technologies for more than a decade as a senior research engineer at Pacific Northwest National Laboratory where he developed low-cost solutions for industrial implementation of friction stir technologies. Working with numerous industrial suppliers around the world, he has introduced cost-efficient solutions for thermal telemetry, demonstrated novel, low-cost tool materials for friction stir spot, implemented high volume production techniques for aluminum tailor-welded blanks, developed new methodologies for joining dissimilar materials, introduced closed-loop thermal control of additive friction stir deposition, and introduced high-speed refill friction stir spot welding. As an active researcher, he received the R&D 100 award in 2011 and again in 2017 the DOE Vehicle Technologies Office Distinguished Achievement award in 2015, and a western region FLC award for technology transfer in 2015.

As a professor, he was recognized by SME as one of the top 20 academics in smart manufacturing in 2021. He actively reviews friction stir-related literature for numerous publications and has documented his work in more than 80 publications and proceedings and six U.S. patents.

Yutaka Sato is a professor in the Department of Materials Processing at Tohoku University, Japan. He earned a Ph.D. in Materials Processing at Tohoku University (2001). His Ph.D. thesis was titled "Microstructural Study on Friction Stir Welds of Aluminum Alloys." He participated in friction stir research of steels at Brigham Young University for a year in 2003. He was a member of Sub-commission III-B WG-B4 at IIW, which is a working group to build international standardization of friction stir spot welding. His work has focused on metallurgical studies of friction stir welding and processing for more than 20 years. He has obtained fundamental knowledge on development of grain structure, texture evolution, joining mechanism, behavior of oxide-layer on surface, and properties–microstructure relationship. He has received a number of awards including the Kihara Award from the Association for Weld Joining Technology Promotion (2008), Prof. Koichi Masubuchi Award from AWS (2009), Murakami Young Researcher Award from the Japan Institute of Metals (2010), Honda Memorial Young Researcher Award (2011), The Japan Institute of Metals and Materials Meritorious Award (2015), and Light Metal Breakthrough Award from the Japan Institute of Light Metals (2017). He has authored or coauthored more than 300 papers in peer-reviewed journals and proceedings.

Piyush Upadhyay is a senior material scientist at Pacific Northwest National Laboratory. He earned his Ph.D. from the University of South Carolina in 2012 in "Boundary Condition Effects on Friction Stir Welding of Aluminum Alloys." For more than a decade, he has been involved in research and development of FSW and allied technologies to join similar and dissimilar materials. Currently, he leads and contributes to projects on friction stir welding and processing of aluminum, magnesium, copper, steels, and polymers. He has received several awards and recognitions including Aid Nepal Scholarship for undergraduate study (2001), Happy House Foundation Research Fellowship at Kathmandu University (2007), the DOE Energy Efficiency& Renewable Energy Recognition for Innovation (2015), and the R&D 100 Award (2018). He has authored or coauthored more than 50 papers in peer-reviewed journals and proceedings and is actively involved in technical societies currently serving as chair of the TMS Shaping and Forming committee and as a guest editor and reviewer for technical journals.

Anton A. Naumov is an associate professor at the Institute of Mechanical Engineering, Materials and Transport at Peter the Great St. Petersburg Polytechnic University (SPbPU). He obtained B.S. (2003), masters (2005), and doctorate (2010) degrees in Metallurgy at SPbPU. In 2012, 2014, and 2016 he won the Grants of the President of Russian Federation for young Ph.Ds in technical field of science. The 2016 grant was focused on the microstructure and mechanical properties evolution of aluminum alloys during friction stir welding. He was one of the organizers for the Laboratory of Lightweight Materials and Structures (LWMS) at SPbPU in 2014 in terms of Mega-Grant of Ministry of Science and Education of Russian Federation for attracting leading foreign scientists and opening R&D labs under their supervision. He has been working in R&D laboratory LWMS until the present time as a principal researcher in the field of friction stir welding and processing. He is an active member of the professional societies TMS and AWS and reviews friction stir related articles for several journals. He has authored or coauthored more than 45 papers in peer-reviewed journals and proceedings and has six Russian Federation patents.

Nilesh Kumar has been an assistant professor in the Department of Metallurgical and Materials Engineering at The University of Alabama since August 2018. Prior to this, he was a post-doctoral research associate in the Department of Nuclear Engineering at North Carolina State University Raleigh and at the University of North Texas Denton. He obtained his Ph.D. degree in Materials Science and Engineering from Missouri University of Science & Technology Rolla. A common theme which cuts across all the research work, he has carried out so far is establishing correlation among processing (thermo-mechanical, friction stir processing, and laser), microstructure (grain-size, dislocations, and precipitates), and mechanical properties (strength, ductility, creep, creep-fatigue, residual stress, and stress corrosion cracking) of metallic materials (aluminum alloys, magnesium alloys, austenitic stainless steels, titanium alloys, high-entropy alloys, advanced high strength steel) used primarily in transportation and power-generation industries. He has published 41 papers in peer-reviewed journals, coauthored 4 books, and contributed 4 book/handbook chapters. He is a member of several scientific organizations and has been a reviewer for more than two dozen scientific journals. He is also the recipient of the Kent D. Peaslee Junior Faculty Award by the Association for Iron & Steel Technology (AIST) Foundation for the year 2019–2020 and was awarded the title "AIST Foundation Steel Professor" in 2022 by the AIST Foundation.

Part I
Additive Friction Stir Deposition

Hybrid Manufacturing: Combining Additive Friction Stir Deposition, Metrology, and Machining

Joshua Kincaid, Ross Zameroski, Timothy No, John Bohling, Brett Compton, and Tony Schmitz

Abstract Aerospace flight panels must provide high strength with low mass. For aluminum panels, it is common practice to begin with a wrought plate and remove the majority of the material to attain the desired structure, comprising a thinner plate with the desired pattern of reinforcement ribs. As an alternative, this study implements hybrid manufacturing, where aluminum is first deposited on a baseplate only at the rib locations using additive friction stir deposition (AFSD). Structured light scanning is then used to measure the printed geometry. This geometry is finally used as the stock model for computer numerical control (CNC) machining. This paper details the hybrid manufacturing process that consists of: AFSD to print the preform, structured light scanning to generate the stock model and tool path, three-axis CNC machining, and post-process measurements for part geometry and microstructure.

Keywords Hybrid manufacturing · Additive friction stir deposition · Structured light scanning · Machining · Microstructure

J. Kincaid · R. Zameroski · J. Bohling · B. Compton · T. Schmitz (✉)
Mechanical, Aerospace, and Biomedical Engineering Department, University of Tennessee, Knoxville, 1512 Middle Drive, Knoxville, TN 37996, USA
e-mail: tony.schmitz@utk.edu

J. Kincaid
e-mail: jkincai4@vols.utk.edu

R. Zameroski
e-mail: rzameros@vols.utk.edu

J. Bohling
e-mail: jbohling@vols.utk.edu

B. Compton
e-mail: bcompto1@utk.edu

T. No · T. Schmitz
Manufacturing Demonstration Facility, Oak Ridge National Laboratory, 2350 Cherahala Blvd, Knoxville, TN 37932, USA
e-mail: tno@vols.utk.edu

Y. Hovanski et al. (eds.), *Friction Stir Welding and Processing XII*,
The Minerals, Metals & Materials Series,
https://doi.org/10.1007/978-3-031-22661-8_1

Introduction

Metal additive manufacturing (AM) includes beam-based technologies, such as powder bed fusion and directed energy deposition, as well as wire arc AM. For these processes, the metal powder or wire is melted using a high-intensity heat source and deposited in a layer-by-layer fashion. The gross part geometry is dictated by the computer model fed to the printer, while the microstructure is established during the subsequent solidification and cooling, ultimately dictated by the local temperature gradient and cooling rate [1]. An alternative non-beam-based, solid-state additive process is provided by additive friction stir deposition (AFSD) [2–5]. In this case, no melting occurs, and the geometry and microstructure are defined by the kinetic energy introduced by the AFSD process. In this sense, AFSD microstructure depends on thermomechanical, rather than solidification, mechanisms. Research efforts have included the study of microstructure and its relationship to mechanical properties and operation parameters [6–15]. Deposition materials have included aluminum, magnesium, copper, and steel alloys [16–20], for example. Repair and cladding [21–24], effect of alloy temper [25], fatigue behavior [26], process modeling [27], and force/temperature control [28] have also been examined in the literature.

The AFSD process is described in Fig. 1. The feedstock is a square metal rod (e.g., wrought material). It is forced through the rotating spindle using a screw-type actuator located above the spindle against the baseplate. The actuator allows rotation of the 0.5 m long feedstock with the spindle while simultaneously providing the axial force and material feed. The feedstock rotation against the baseplate generates frictional heat, which softens the feedstock sufficiently to cause plastic flow and, ideally, a metallurgical bond with the existing material. The printed material is constrained axially by the gap between the rotating tool and baseplate (1–3 mm). In the lateral direction, there is only friction between the plastically flowing material and the tool (on the top) and baseplate (on the bottom). For this reason, flash can occur at the outer portions of the layer. The tool is translated parallel to the base plate with a selected feed rate to print the layer. For subsequent layers, material is deposited on the previous layer so the microstructure for the previous layer is affected by the plastic deposition from the new layer.

Hybrid Manufacturing Steps

To approximate an aerospace flight panel, a ribbed structure with a center hole and boss was designed; see Fig. 2 (left panel), where the ribs are 12.7 mm wide and 10.2 mm tall. The preform was fabricated using a MELD Manufacturing L3 machine to deposit 6061-T6 aluminum square rod with a side length of 9.5 mm onto a 6061-T6 aluminum baseplate. To enable deposition, the AFSD tool paths were generated. These included four overlapping C-shaped paths and a central cylinder that were printed in five layers. The total thickness of the five-layer deposition was 12.7 mm.

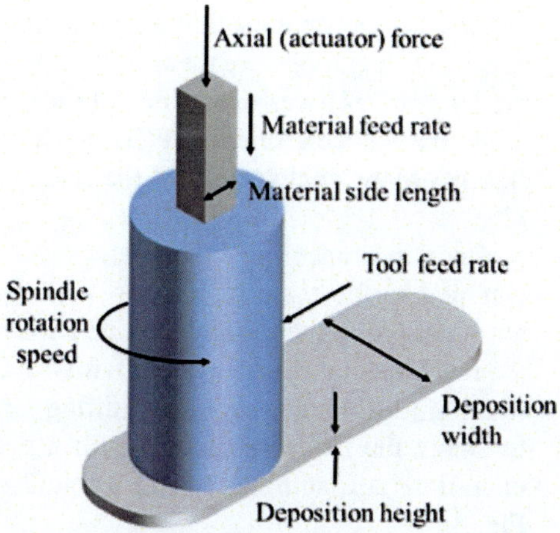

Fig. 1 AFSD description

The preform is displayed in Fig. 2 (middle panel) and the operating parameters are provided in Table 1.

The printed preform was measured using a GOM ATOS Q structured light scanner. The scan model was imported into computer-aided manufacturing (CAM) software, where it was used as the stock model for toolpath generation. The scan model was aligned with the computer-aided design (CAD) model of the ribbed panel, and the coordinate system was assigned using the corner of the baseplate; see Fig. 2 (right).

Fig. 2 (Left) CAD model of panel. (Middle) 6061 aluminum preform. (Right) scanned preform model

Table 1 AFSD operating parameters

Layer	Start spindle speed (rpm)	Final spindle speed (rpm)	Tool feed rate (mm/min)	Material feed rate (mm/min)
1	300	275	102	152
2	300	275	102	152
3	275	275	152	168
4	275	250	142	183
5	275	225	142	198

To select optimal machining parameters, the frequency response functions (FRFs) of each cutting tool were measured using impact testing. Here, a modal hammer (PCB model 086C04) was used to excite the tool tip and the response was measured by a low-mass accelerometer (PCB model 352C23). The FRFs were used to generate stability maps, which enabled the selection of optimal, stable machining parameters [29].

Once the machining parameters and toolpaths were selected, the preform was clamped to the table of a Haas VF-4 three-axis CNC milling machine. The part was then probed with the machine's touch trigger probe to locate the part and align the machine coordinate system with the CAM coordinate system using a coordinate rotation [30]. Facing, contour milling, and boring operations were all implemented to create the ribbed structure with a hole and boss in the center. The machining operations are summarized in Table 2, and the CAM toolpaths are displayed in Fig. 3.

Table 2 CNC milling operations and parameters

Operation	Tool	Spindle speed (rpm)	Feed (mm/min)
1. Face the top surface	76.2 mm diameter, 8 insert facemill	5000	2032
2. Rough inner pocket	19.05 mm diameter, 3 flute endmill	7000	1600.2
3. Finish inner pocket	6.35 mm diameter, 3 flute endmill	7700	320.0
4. Mill center hole	19.05 mm diameter, 3 flute endmill	7000	1600.2
5. Rough outer profile	19.05 mm diameter, 3 flute endmill	7000	1600.2
6. Finish outer profile	6.35 mm diameter, 3 flute endmill	7700	320.0

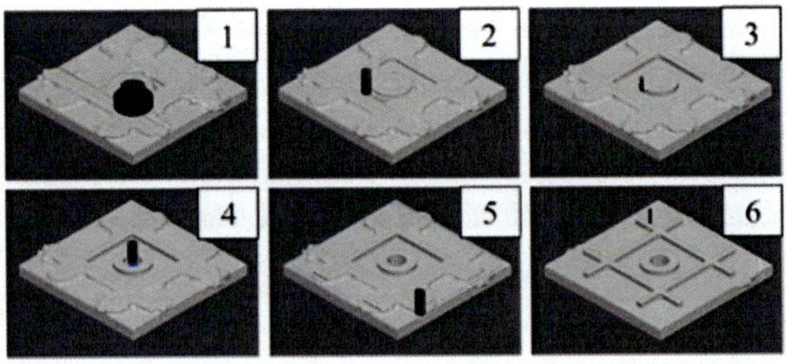

Fig. 3 CNC milling operations from Table 2

Results and Discussion

Each step in the hybrid manufacturing process was completed with the expected results. The AFSD toolpaths deposited material in the desired locations on the baseplate. The structured light scan provided a stock model for the CAM software that was used to set a coordinate system and define the toolpaths required to remove the excess material. Stable machining conditions were observed using the parameters obtained from tap testing. In summary, the structured light scanning strategy to provide a CAM stock model and local coordinate system was successfully implemented.

One advantage of the hybrid manufacturing approach is reduced material use and removal (by machining). If the part had been machined from wrought plate stock, the required starting volume would have been 3539 cm^3. In comparison, the preform volume (i.e., the baseplate and printed material) was 3163 cm^3. This represents a material savings of 376 cm^3, or 10.6%, by the hybrid approach. Note that the baseplate comprised 2360 cm^3 of this total volume. Neglecting the baseplate, the "top" wrought volume that contained the ribs and boss was 1180 cm^3 and the printed volume was 804 cm^3. The 376 cm^3 volume savings is now 31.9% of the "top" wrought volume. If machining from the full wrought plate, the amount of material to be removed would have been 1090 cm^3 to obtain the final geometry. Machining the printed preform, on the other hand, required 714 cm^3 to be removed. The difference is again 376 cm^3 for a reduction in material removal volume of 34.5%. This is similar to the advantages obtained by wire arc AM in aluminum [31].

To quantify the geometric fidelity of the machined part, measurements were completed using the structured light scanner. The scan results were compared to the CAD model by creating a best-fit alignment and observing the differences; see Fig. 4. The alignment showed a maximum deviation of 0.25 mm from the CAD model; see Fig. 5, where red indicates extra material and blue less material than desired. One potential cause of these deviations is the release of internal stresses during the machining process. In follow-on testing, measurements will be performed before deposition, after deposition, and after machining to record any part distortion.

Finally, sections were cut from the machined part by water jet to characterize the as-deposited microstructure obtained from the AFSD process. Two sections were studied. These included the overlap region of two C-shaped paths (labeled tracks 1 and 2 in Fig. 6) and a standalone region from the middle of track 1. In Fig. 6, track

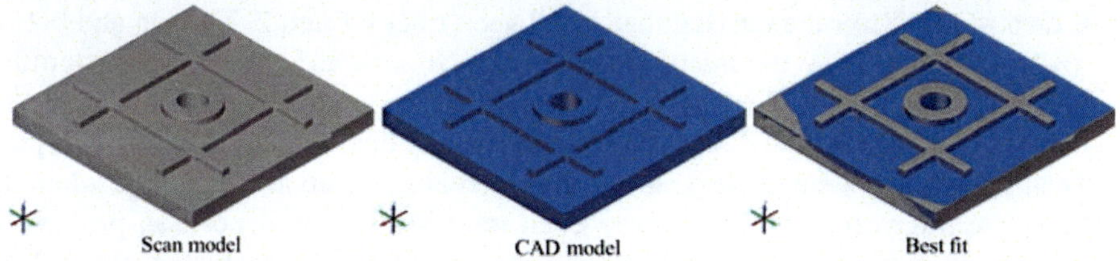

Fig. 4 Comparison of scan and CAD models to determine part errors

Fig. 5 Difference map between CAD and scan models

1 (red arrows) was deposited first and track 2 (white arrows) was deposited second; the circular arrows indicate the tool rotation direction for each track. The overlap sample (white box) extended across the full width of the overlap region between the two tracks and was oriented normal to the overlap interface. The standalone sample (red box) extended across the width of the track 1 and was oriented normal to the tool feed direction. Both samples were polished and then etched with Weck's reagent for optical microscopy.

The etched cross section of the overlap region is displayed in Fig. 7. In this figure, the baseplate region at the bottom, track 1, and track 2 are identified. In addition, the build direction (i.e., bottom to top layers) and the advancing side (AS) and retreating side (RS) of the deposition are labeled. The advancing side occurs where the peripheral velocity of the rotating tool adds to the tool feed rate, while the retreating side occurs where the peripheral velocity is opposite the tool's feed direction. In Fig. 7, three zones (A, B, and C) are highlighted for higher magnification images and the boundary between tracks 1 and 2 is shown as a dashed line as a guide to the eye.

Optical micrographs for zones A, B, and C, as well as the wrought baseplate are shown in Fig. 8 for the overlap region between tracks 1 and 2. The top and bottom rows of images show the microstructure of each zone at low and high magnification, respectively. It is observed that the baseplate microstructure exhibits elongated features in the horizontal direction due to the rolling operation used to form the plate. The interface between the deposition and baseplate is seen at location A, where the baseplate microstructure has clearly been modified by the deposition process. At location B, a wedge-shaped feature is seen due to the interaction of track 2 and

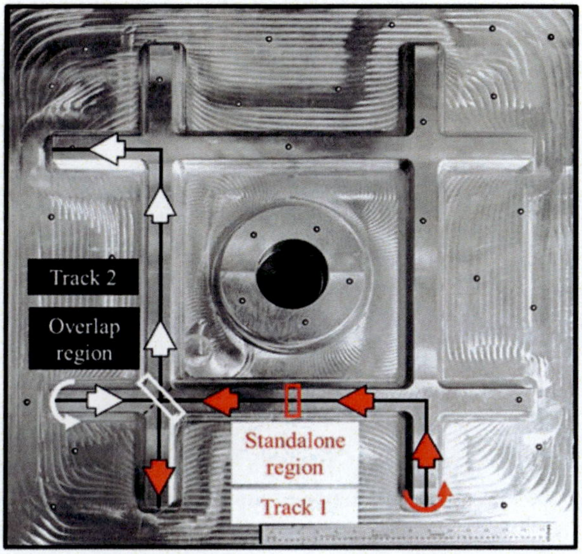

Fig. 6 Standalone (track 1) and overlap (tracks 1 and 2) regions for microstructure evaluation

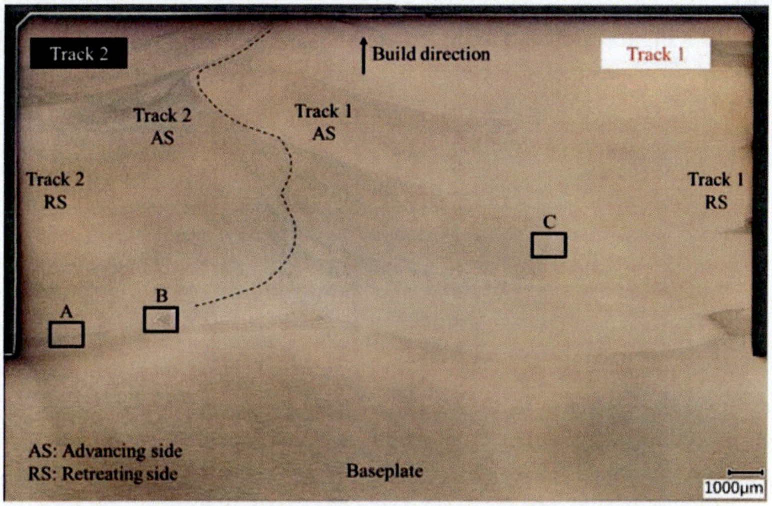

Fig. 7 Overlap region (tracks 1 and 2) microstructure

baseplate. Location C shows a uniform, fine grain microstructure in the deposited material due to the severe plastic deformation for the AFSD process.

The etched cross section of the standalone (track 1) region is shown in Fig. 9. In this figure, the baseplate region, deposit region, and advancing and retreating sides are labeled; the tool feed direction proceeded out of the image plane during deposition.

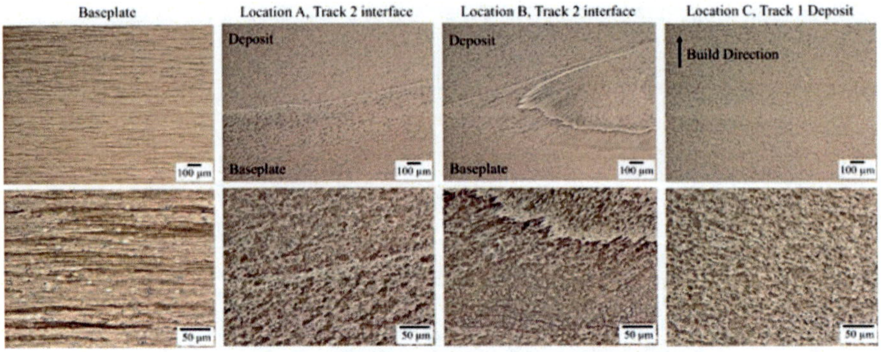

Fig. 8 Individual images for zones in overlap region (tracks 1 and 2)

Optical micrographs for zones A, B, and C, as well as the wrought baseplate are shown in Fig. 10. The interface between the deposition and baseplate is shown for location A, where the baseplate microstructure is distinctly different before and after deposition. At location B, a wedge-shaped feature is again seen at the baseplate-deposit interface. Location C shows a uniform, fine grain microstructure for the deposited material. Further characterization of the microstructural changes in the base plate, interface features observed in both regions (Figs. 8 and 10), and hardness properties will be conducted to understand the effect on behavior of the final part.

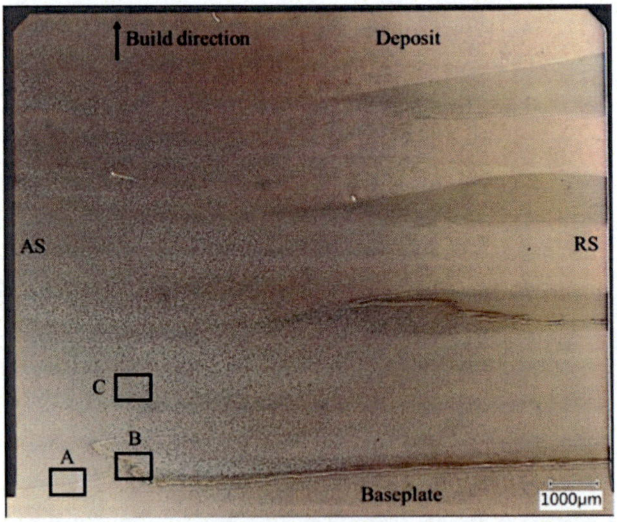

Fig. 9 Standalone region (track 1) microstructure

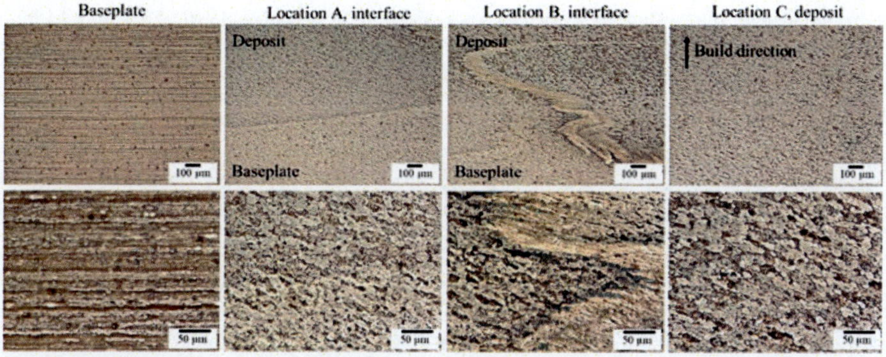

Fig. 10 Individual images for zones in standalone region (track 1)

Conclusions

This paper described the combination of AFSD, structured light scanning, and CNC machining in a hybrid manufacturing scenario. The demonstration part was an aluminum aerospace flight panel, although the material and application are not limited to this domain. The AFSD material demonstrated machinability similar to wrought aluminum. The structured light scanning procedure provided an accurate stock model for the CAM software, while also ensuring the desired part geometry was contained within the printed preform. Tool tip FRF measurements enabled the selection of optimal machining parameters. Post-process measurements were used to compare the final part with the intended CAD design and evaluate the microstructure of the baseplate and deposited material. Ultimately, this work demonstrated a hybrid manufacturing approach that leverages AFSD, metrology, and machining to provide a new option for the production of aerospace flight panels, as well as other metallic components traditionally obtained from wrought plate, castings, or forgings.

References

1. Dehoff RR, Kirka MM, Sames WJ, Bilheux H, Tremsin AS, Lowe LE, Babu SS (2015) Site specific control of crystallographic grain orientation through electron beam additive manufacturing. Mater Sci Technol 31(8):931–938
2. Hang ZY, Jones ME, Brady GW, Griffiths RJ, Garcia D, Rauch HA, Cox CD, Hardwick N (2018) Non-beam-based metal additive manufacturing enabled by additive friction stir deposition. Scripta Mater 153:122–130
3. Khodabakhshi F, Gerlich AP (2018) Potentials and strategies of solid-state additive friction-stir manufacturing technology: a critical review. J Manuf Process 36:77–92
4. Yu HZ, Mishra RS (2021) Additive friction stir deposition: a deformation processing route to metal additive manufacturing. Mater Res Lett 9(2):71–83
5. Gopan V, Wins KLD, Surendran A (2021) Innovative potential of additive friction stir deposition among current laser based metal additive manufacturing processes: a review. CIRP J Manuf Sci Technol 32:228–248

6. Priedeman JL, Phillips BJ, Lopez JJ, Tucker Roper BE, Hornbuckle BC, Darling KA, Jordon JB, Allison PG, Thompson GB (2020) Microstructure development in additive friction stir-deposited Cu. Metals 10(11):1538
7. Perry ME, Griffiths RJ, Garcia D, Sietins JM, Zhu Y, Hang ZY (2020) Morphological and microstructural investigation of the non-planar interface formed in solid-state metal additive manufacturing by additive friction stir deposition. Addit Manuf 35:101293
8. Griffiths RJ, Garcia D, Song J, Vasudevan VK, Steiner MA, Cai W, Hang ZY (2021) Solid-state additive manufacturing of aluminum and copper using additive friction stir deposition: process-microstructure linkages. Materialia 15:100967
9. Agrawal P, Haridas RS, Yadav S, Thapliyal S, Gaddam S, Verma R, Mishra RS (2021) Processing-structure-property correlation in additive friction stir deposited Ti-6Al-4V alloy from recycled metal chips. Addit Manuf 47:102259
10. Phillips BJ, Mason CJT, Beck SC, Avery DZ, Doherty KJ, Allison PG, Jordon JB (2021) Effect of parallel deposition path and interface material flow on resulting microstructure and tensile behavior of Al-Mg-Si alloy fabricated by additive friction stir deposition. J Mater Process Technol 295:117169
11. Perry ME, Rauch HA, Griffiths RJ, Garcia D, Sietins JM, Zhu Y, Zhu Y, Hang ZY (2021) Tracing plastic deformation path and concurrent grain refinement during additive friction stir deposition. Materialia 18:101159
12. Williams MB, Robinson TW, Williamson CJ, Kinser RP, Ashmore NA, Allison PG, Jordon JB (2021) Elucidating the effect of additive friction stir deposition on the resulting microstructure and mechanical properties of magnesium alloy we43. Metals 11(11):1739
13. Mukhopadhyay A, Saha P (2022) A critical review on process metrics–microstructural evolution–process performance correlation in additive friction stir deposition (AFS-D). J Braz Soc Mech Sci Eng 44(9):1–35
14. Joshi SS, Sharma S, Radhakrishnan M, Pantawane MV, Patil SM, Jin Y, Yang T, Riley DA, Banerjee R, Dahotre NB (2022) A multi modal approach to microstructure evolution and mechanical response of additive friction stir deposited AZ31B Mg alloy. Sci Rep 12(1):1–15
15. Zeng C, Ghadimi H, Ding H, Nemati S, Garbie A, Raush J, Guo S (2022) Microstructure evolution of Al6061 alloy made by additive friction stir deposition. Materials 15(10):3676
16. Phillips BJ, Avery DZ, Liu T, Rodriguez OL, Mason CJT, Jordon JB, Brewer LN, Allison PG (2019) Microstructure-deformation relationship of additive friction stir-deposition Al–Mg–Si. Materialia 7:100387
17. Alzahrani B, El-Sayed Seleman MM, Ahmed MM, Elfishawy E, Ahmed AM, Touileb K, Jouini N, Habba MI (2021) The applicability of die cast A356 alloy to additive friction stir deposition at various feeding speeds. Materials 14(20):6018
18. Joshi SS, Patil SM, Mazumder S, Sharma S, Riley DA, Dowden S, Banerjee R, Dahotre NB (2022) Additive friction stir deposition of AZ31B magnesium alloy. J Magnesium Alloys
19. Garcia D, Hartley WD, Rauch HA, Griffiths RJ, Wang R, Kong ZJ, Zhu Y, Hang ZY (2020) In situ investigation into temperature evolution and heat generation during additive friction stir deposition: a comparative study of Cu and Al-Mg-Si. Addit Manuf 34:101386
20. Beladi H, Farabi E, Hodgson PD, Barnett MR, Rohrer GS, Fabijanic D (2022) Microstructure evolution of 316L stainless steel during solid-state additive friction stir deposition. Phil Mag 102(7):618–633
21. Griffiths RJ, Petersen DT, Garcia D, Yu HZ (2019) Additive friction stir-enabled solid-state additive manufacturing for the repair of 7075 aluminum alloy. Appl Sci 9(17):3486
22. Hartley WD, Garcia D, Yoder JK, Poczatek E, Forsmark JH, Luckey SG, Dillard DA, Hang ZY (2021) Solid-state cladding on thin automotive sheet metals enabled by additive friction stir deposition. J Mater Process Technol 291:117045
23. Avery DZ, Cleek CE, Phillips BJ, Rekha MY, Kinser RP, Rao HM, Brewer LN, Allison PG, Jordon JB (2022) Evaluation of microstructure and mechanical properties of Al-Zn-Mg-Cu alloy repaired via additive friction stir deposition. J Eng Mater Technol 144(3):031003
24. Peter Martin L, Luccitti A, Walluk M (2022) Evaluation of additive friction stir deposition for the repair of cast Al-1.4 Si-1.1 Cu-1.5 Mg-2.1 Zn. J Manuf Sci Eng 144(6)

25. Ahmed MM, El-Sayed Seleman MM, Elfishawy E, Alzahrani B, Touileb K, Habba MI (2021) The effect of temper condition and feeding speed on the additive manufacturing of AA2011 parts using friction stir deposition. Materials 14(21):6396
26. Anderson-Wedge K, Avery DZ, Daniewicz SR, Sowards JW, Allison PG, Jordon JB, Amaro RL (2021) Characterization of the fatigue behavior of additive friction stir-deposition AA2219. Int J Fatigue 142:105951
27. Stubblefield GG, Fraser K, Phillips BJ, Jordon JB, Allison PG (2021) A meshfree computational framework for the numerical simulation of the solid-state additive manufacturing process, additive friction stir-deposition (AFS-D). Mater Des 202:109514
28. Merritt GR, Williams MB, Allison PG, Jordon JB, Rushing TW, Cousin CA (2022) Closed-loop temperature and force control of additive friction stir deposition. J Manuf Mater Process 6(5):92
29. Schmitz T, Smith KS (2019) Machining dynamics: frequency response to improved productivity, 2nd edn. Springer, New York
30. Cornelius A, Dvorak J, Jacobs L, Penney J, Schmitz T (2021) Combination of structured light scanning and external fiducials for coordinate system transfer in hybrid manufacturing. J Manuf Process 68:1824–1836
31. Panchenko O, Kurushkin D, Mushnikov I, Khismatullin A, Popovich A (2020) A high-performance WAAM process for Al–Mg–Mn using controlled short-circuiting metal transfer at increased wire feed rate and increased travel speed. Mater Des 195:109040

Closed-Loop PID Temperature Control of Additive Friction Stir Deposition

Jason Glenn, Luk Dean, Arnold Wright, and Yuri Hovanski

Abstract Additive friction stir deposition (AFSD), a derivative technology of friction stir welding (FSW), is now rapidly growing in industries that require bulk additive manufacturing. For mass adoption to occur, deposition consistency and quality along with thermal inputs and their impacts need to be better understood and improved. In both AFSD and FSW, it has been demonstrated that properties of depositions or welds vary along their length in response to in-process temperatures. Research using temperature control to maintain weld temperatures in FSW has been performed for some years now with significant results in both quality and reliability, but this technology is just starting to be applied toward AFSD. This work describes the implementation of a temperature control method that was proven successful in FSW to AFSD. It demonstrates the accuracy achievable using that method for AA 7050 T7451 compares the controlled results against a fixed RPM deposition, reviews results of temperature control over multi-pass AFSD, and outlines future work for AFSD involving temperature control.

Keywords AFSD · Additive friction stir deposition · Temperature control · PID control

J. Glenn (✉) · L. Dean · Y. Hovanski
Brigham Young University, 265 CTB, Provo, UT, USA

L. Dean
e-mail: ldean3@byu.edu

Y. Hovanski
e-mail: yuri.hovanski@byu.edu

A. Wright
Bond Technologies, 1353 Wade Dr. Ste B, Elkhart, IN, USA

Y. Hovanski et al. (eds.), *Friction Stir Welding and Processing XII*,
The Minerals, Metals & Materials Series,
https://doi.org/10.1007/978-3-031-22661-8_2

Introduction

Additive friction stir deposition (AFSD) is a bulk additive manufacturing process based on friction stir welding (FSW). Instead of the rotating tool head having a pin that protrudes into the substrate like in FSW, there is a hole in the middle through which filler material may be extruded. Filler material has taken the form of powder, square rods, and even recycled chips [1], but it is most commonly performed using square rods. The material extruding out of the tool is plastically deformed, filling an area in between the tool shoulder and the substrate. As linear motion is introduced, strips of material are deposited ranging from 5 to 40 mm wide depending on the shoulder diameter and 1–5 mm thick depending on the specified settings. The process can then lift in the Z-axis and deposit layer upon layer until the desired near-net shape (build) is produced. The rotating shoulder both ensures adequate mixing between layers and shears grains resulting in fully dense parts with a fine equiaxed crystal structure [2].

Despite the advantages to AFSD, there are still many areas that need further research. One study found that material properties in a precipitation hardened material varied over each deposition layer in a 38 mm stack of 1 mm layers [3]. While the top layers had an as-deposited yield strength (YS) of approximately 220 MPa, the bottom layer was reduced to about 110 MPa. Similar issues can be seen in FSW where properties in precipitation hardened materials can vary even along the length of a single weld [4]. These issues were demonstrated as problems arising from weld temperature in a study that varied weld backing plates to increase heat transfer away from the weld [5]. Aluminum backing plates with high thermal conductivity increased weld properties by maintaining a lower weld temperature while low thermal conductivity backing plates such as ceramic and steel reduced post weld properties.

To approach this problem in FSW, an increasing number of groups are turning to temperature-controlled welds. There are a number of temperature control methods available including PID, cascading PID RPM loops [6, 7], cascading PID torque loops [8], and more. Wright et al. reviewed different methods of temperature control including the previously listed methodologies and found that cascading PID loops commanding either RPM or torque was the most robust controllers enabling temperature control within ±3 °C. Additionally, while not as robust, a simple PID loop commanding RPM was shown to be more easily implemented on an average industry machine, while maintaining ±5 °C [9]. Though the simple loop is not as robust as the other methods, it was deemed sufficient for the needs of this experiment. Due to its ease of implementation, the authors chose to use this method for this work. Research performed in FSW using PID RPM control found that post-weld properties could be quickly optimized and the standard deviation for properties along the weld could be minimized [10]. Similar capabilities are also needed in AFSD along with an understanding and control of thermal inputs for residual stress management. Temperature control is just beginning to be implemented in AFSD [11] so such results do not yet exist. AFSD also poses further issues with the implementation of temperature control in that layers stack onto each other resulting in vastly different thermal boundary conditions from one layer to the next. It is unsure whether such a multi-pass system can still be adequately controlled without changing control parameters.

The purpose of this current study was to evaluate the efficacy of a PID spindle speed temperature control method, commonly used for FSW, in AFSD as a means of increasing the consistency of thermal input into additive depositions. A comparison of controlled and non-controlled AFSD deposits is examined to determine its success of implementation. Then, temperature control data from 4 stacks consisting of 7 layers in height using varied PID values (the kp, ki, and kd constants in the equations) are compared to see if adjustments in the control system are necessary to compensate for changing boundary conditions. An outline for further development of temperature-controlled AFSD is also given.

Methodology

Implementation Depositions

Two depositions were performed each 125 mm in length. The linear traverse speed was 150 mm/min, and the material extrusion rate was 4770 mm^3/min. The first deposition was performed at a fixed spindle speed of 400 RPM which is comparable to other published papers when correlating surface speed of the tool radii. The second was performed using PID RPM control. Feed rate and linear velocity were held constant for both deposits, and no anvil cooling was applied meaning anvil temperatures increase, though only slightly due to the high thermal mass of the anvil, along the length of the deposition.

Multi-Pass Control Depositions

In this experiment, 28 total depositions were performed, each 150 mm in length. The depositions consisted of 4 stacks of 7 each performed on individual 150 mm × 200 mm × 6.25 mm plates. A thermocouple (TC 1 Fig. 1) was inserted into the bottom of the plate, centered, so the deposition would pass over halfway through the deposition. Another thermocouple (TC 2 Fig. 1) was placed on the anvil approximately 100 mm from the edge of the substrate. The first deposition on each substrate was started while the substrate and anvil were 25 ± 2 °C. Layers 2–7 for each of the plates were performed consecutively, each starting two minutes after the end of the previous. The two minutes gave sufficient time for recording and updating PID values between each layer and for the feed rod swap that was necessary before layer five of each stack. The first stack of seven consisted of seven tuning runs to record the change in tuned PID value outputs corresponding to the different boundary conditions. The second stack of seven was performed using temperature control with the

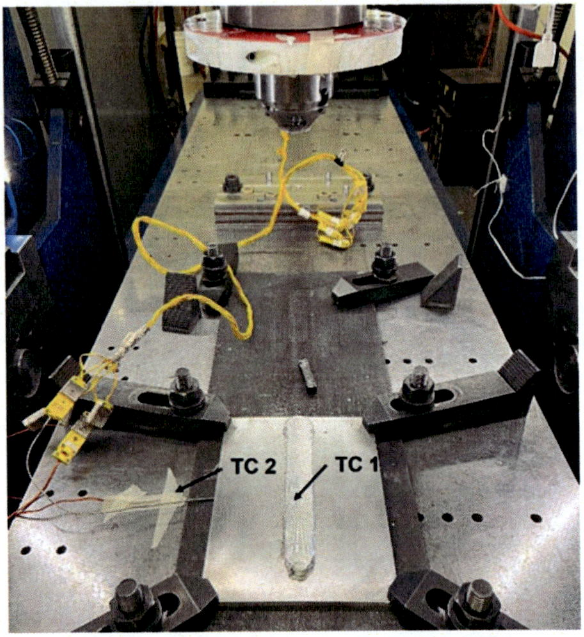

Fig. 1 Image of the experiment set up for a 7-layer stack. Note thermocouple one placed under the deposition path and thermocouple two placed 100 mm away from deposition path on the anvil

PID values constant as the first layer tuning run output from the previous stack. The third stack was performed using temperature control while updating the PID values for each layer to reflect the outputs from each corresponding layer of stack one tuning depositions. The fourth stack was performed using temperature control with a fixed PID values that were the average of all seven layers outputs from the 1st stack. Feed rate and linear velocity were held constant for all deposits.

Temperature Building

From experimentation while performing hundreds of depositions, three temperature building phases of a deposition have been discovered. The first is from frictional rubbing, under a high load, of the shoulder against the substrate. This is performed while monitoring the tool temperature until it reaches a specific material dependent temperature at which the material may be extruded. At this point, the deposition program moves on to phase two and begins extruding as it lifts up to the specified layer height. The third phase is when linear motion is enabled and begins the moment the tool reaches the correct deposition height. Each of these phases are responsible for significant portions of the temperature build in a deposition and as such have had to be controlled.

In phase one, an alloy specific RPM and load are specified and then a program is implemented that monitors the tool temperature and continues the rubbing until it reaches an extrudable temperature.

During phase two, extrusion rates are varied using a fixed high and fixed low value depending on temperature so that excessive temperature gain or any temperature reduction are avoided. This phase generally aims for some high fraction of the temperature set point (approximately 80% for AA7050). The value is determined by experimentation as its best to approach the set point but to leave some room the temperature increases that will come from phase three.

In phase three, a conditional controller either increases RPM while the tool temperature is below the temperature set point or decrease RPM when the tool temperature is approaching or above the set point. This occurs during a 15 mm ramp, and the goal is to keep the temperature to within ± 10 °C of the temperature set point by the end. This allows autotuning runs to calculate gains more representative of steady-state AFSD rather than of the startup phase. It also reduces the 'distance to control' of temperature-controlled runs and avoids large RPM steps that can negatively affect the deposition. On the authors' machine, these phases are all coded into the deposition program, and therefore, they perform automatically. This results in a successful deposition every time with the push of a button.

Machine and Materials

AFSD deposits were performed on a TTI model RM-2 Linear FSW machine with an RPT axis integrated by BOND Technologies. All deposits were performed using AA7050-T7451 for both the feed material and the substrates. The feed rods are 9.27 mm $\times$ 9.27 mm bars cut from 9.525 mm plate via waterjet and milled to the correct dimensions. The substrates are 152 mm $\times$ 200 mm $\times$ 6.35 mm in dimension. The deposits are centered on the substrate plate, with the 125 mm long direction aligned with the 200 mm substrate length.

Temperature Measurement

The temperature of the tooling was measured with a type-K thermocouple (OMEGA #KMQSS-062G-6), with the tip positioned close to the tool surface as shown in Fig. 2. The thermocouple was connected to a measurement and Bluetooth transmitter board (LORD #TC-Link-200-OEM) in a custom enclosure attached to the tooling. This transmitted to a receiver device (LORD #WSDA-Base-101-LXRS) that provided an analog 0–3.3 V DC signal into an input card connected to the PLC. The temperature was updated by the measurement board at a frequency of 32 Hz.

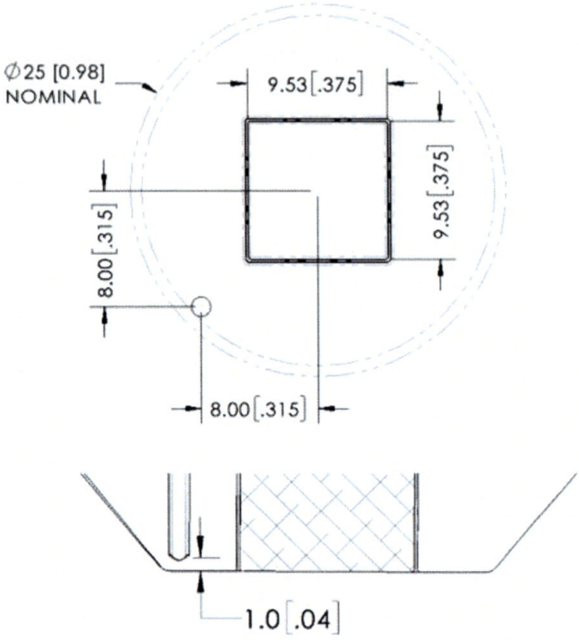

Fig. 2 Location of the thermocouple in the tool

Temperature Control Method

The author chose single-loop proportional-integral-derivative (PID) control with the thermocouple temperature as the input and the spindle speed as the output. The control scheme used also has an anti-windup feature that is active until the temperature first reaches within 5 °C of the temperature set point. During the anti-windup phase, the integral gain is set as 0, while the P & D values are the values calculated by the auto-tuner. Once the temperature has reached within 5 °C of the set point, the integral gain is set to the auto-tuned value and remains that value until the end of the deposition, regardless of the error in temperature. If the PID gains are not appropriate for the process, it may become unstable, saturating the spindle RPM.

The auto-tuner facility is based on an adaptive relay test as described by Taysom and Sorensen [12]. This tuner was originally developed for use in FSW, which can be modeled as a first-order plus dead-time system. The adaptive relay test uses a series of step changes in spindle speed to perform automatic system identification by evaluating the temperature response.

Use of the Auto-tuner

The authors first developed parameters for a non-temperature-controlled deposition process using position and constant RPM control. Once a manual program can provide close to the desired output (a stable deposition with temperature ±60 °C of the desired set point), the tuning program can be implemented. The tuning program derives P, I, and D values from observing the response time of the measured temperature on the tool face to an applied spindle speed step change. The program is set with a time delay estimate of one second and runs eight cycles (a cycle is considered crossing the temperature set point in response to the spindle speed step changes). For most effective control, the autotuning should be conducted on a section of deposition that is representative of the weld that will be temperature controlled.

Results and Discussion

Implementation

The deposition, Fig. 3, performed with constant spindle speed experiences a temperature climb of approximately 17 °C over 125 mm as shown in Fig. 4. The trajectory of the temperature indicates that if the length of the deposition had been increased the temperature would have continued to climb. This means that for precipitate-strengthened materials like AA7050-T7451 varied properties would be expected throughout the length of the deposition. Though a portion of this temperature climb could be from a lack of anvil cooling, significant temperature increases are observed when extrusion starts and when linear motion starts. The high RPMs necessary to achieve temperatures at which extrusion may commence are excessive compared to the steady-state AFSD rotational speed requirements. Active cooling of the substrate/anvil would just as likely increase the RPM requirements for startup as they would the RPM requirements for steady-state deposition. The result would still likely leave a parameter gap, though differed.

The temperature-controlled deposition, Fig. 5, by comparison quickly reaches the temperature set point of 375 °C and then fluctuates within 0.5 °C under and 1.5 °C above as shown in Fig. 6. Temperature was held within this range from approximately 20 mm into the weld (temperature control was turned on at 15 mm) until the end of the deposition. The spindles rotational speed decreased from approximately 460 RPM at starting to around 220RPM at the end of the weld. The trajectory suggests an optimal steady-state value near 220RPM for these boundary conditions.

Fig. 3 Picture of the fixed 400 RPM deposition

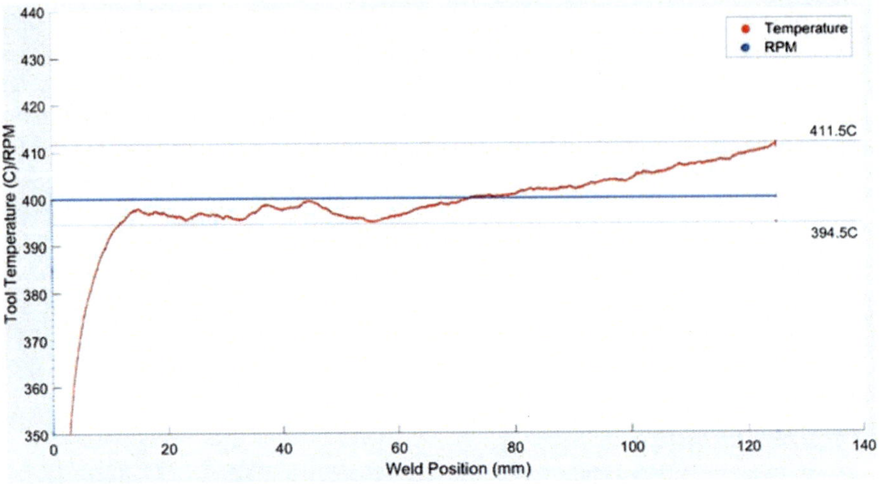

Fig. 4 Temperature and RPM profile of constant RPM AFSD deposit

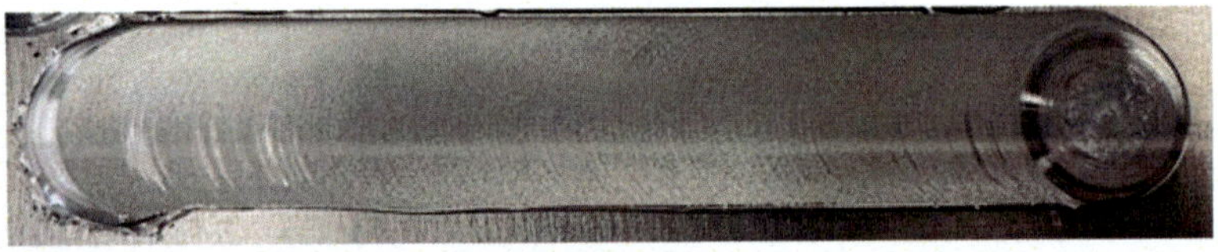

Fig. 5 Picture of temperature-controlled deposition

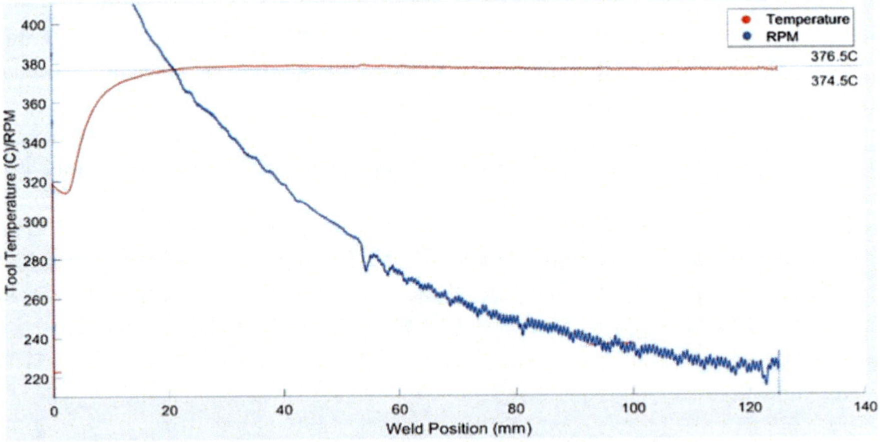

Fig. 6 Temperature and RPM profile of temperature-controlled AFSD deposit

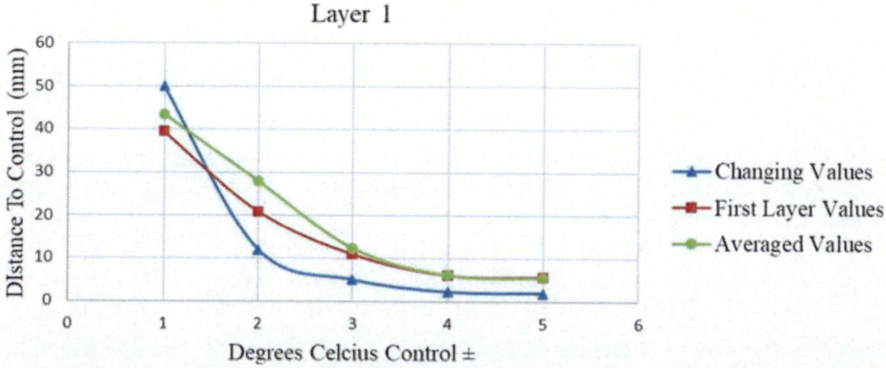

Fig. 7 Comparison plot of layer one of the three different controlled stacks. The vertical axis represents how far along the weld from where temperature control was enabled until control was reached. The horizontal axis represents degrees of control

Multi-pass Temperature Control

Displayed in Figs. 7 and 8 are two plots comparing the distance along the deposition after enabling temperature control against the range of control achieved. The best results were seen in the stacks where the PID values changed with each corresponding layer and where the PID values were fixed from the first layer tuning run. Every deposition achieved a control of ±5 °C within 6 mm of travel after enabling temperature control regardless of how the three different PID values were obtained, (i.e., first layer, average of each layer, or layer by layer). When comparing layer one, Fig. 7, to layer seven, Fig. 8, its apparent that control improves for higher layers. This is likely due to the reduction of heat transfer out as the anvil and substrate warm up resulting in a more stable system. While not shown herein, this trend is also apparent when comparing the plots from layer one to layers two through six.

Evidence from this data leads to the conclusion that control PID values obtained from the first layer are adequate for control at higher layers or on hotter substrates. Despite the changing boundary conditions, sufficient control is obtained. Furthermore, there appears to be little benefit in changing PID values for each layer. Though there may be minor benefits from changing the PID values with each layer, the evidence is not substantial enough to make any claims nor are the results significant enough to be worth the effort.

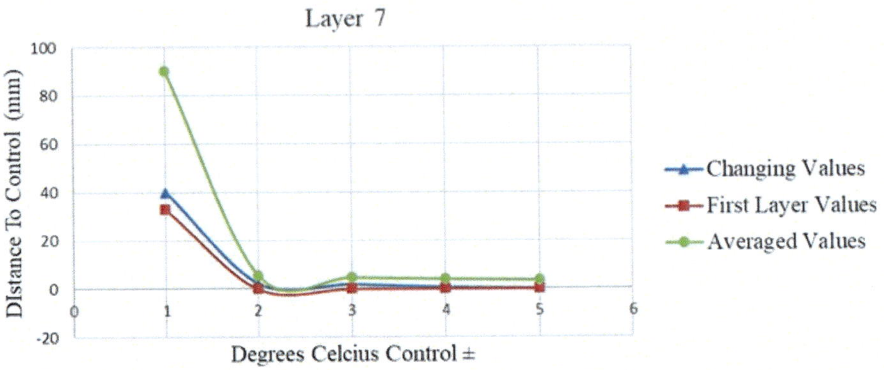

Fig. 8 Comparison plot of layer seven of the three different controlled stacks. The vertical axis represents how far along the weld from where temperature control was enabled until control was reached. The horizontal axis represents degrees of control

Conclusion

Single-loop PID control of the spindle speed in AFSD is indeed a viable option to control thermal input for precipitate-strengthened aluminums such as AA7050-T7451. The consistency in the temperature indicates a likely corresponding consistency of properties along the length of the deposition as demonstrated in FSW [10].

The changing boundary conditions of AFSD from its multi-pass nature do not require changing control parameters to maintain temperature control. Control within ±5 °C is obtained within 6 mm of deposition length, and control of ±2 °C is obtained within 25 mm on the first layer and within 5 mm on subsequent layers. Control with first layer PID values is sufficient and does not appear to improve with individually tuned values. Distance to control is also reduced as the surroundings increase in temperature.

Further Work

Knowing that temperature control can be used on AFSD, work demonstrating that the consistency of deposition temperature, enabled by temperature control, reduces post-deposition properties variation along the length of the deposit. In addition, deposits could be made at varied temperatures and examined to find if a specific deposition temperature provides optimized post-deposition strength.

Furthermore, a model should be developed for AFSD of large builds. Varied thermal inputs and uneven cooling can lead to dramatic residual stress, a phenomenon observed by companies testing samples they had printed via AFSD. Some parts were reported to have serious warping and even have cracked during post-machining heat treatments. This is likely due to residual stress build up from the printing process. These residual stresses can be seen in substrates that had multiple layers printed on them, leaving a visible cambering of the plate. Temperature control will provide valuable data and a consistent variable for this modeling process.

AFSD requires continued work as it is implemented into industry. An understanding of post-deposition properties and resulting residual stresses needs to be obtained before prints of any geometry can be reliably made. Temperature control provides vital data toward achieving those goals and should be implemented more broadly to speed up current research efforts.

References

1. Jordon JB, Allison PG, Phillips BJ, Avery DZ, Kinser RP, Brewer LN, Cox C, Doherty K (2020) Direct recycling of machine chips through a novel solid-state additive manufacturing process. Mater Des 193
2. Yu HZ, Mishra RS (2021) Additive friction stir deposition: a deformation processing route to metal additive manufacturing. Mater Res Lett 9(2):71–83
3. Mason CJT, Rodriguez RI, Avery DZ, Phillips BJ, Bernarding BP, Williams MB, Cobbs SD, Jordon JB, Allison PG (2021) Process-structure-property relations for as-deposited solid-state additively manufactured high-strength aluminum alloy. Additive Manuf 40
4. Ugla AA (2011) Experimental investigation of variation in mechanical and microstructural properties along the length of similar AA6063 FSW. Thi_Qar Univ J Eng Sci 2(4):17
5. Upadhyay P, Reynolds A (2014) Effect of backing plate thermal property on friction stir welding of 25-mm-Thick AA6061. Metall Mater Trans A 45:2091–2100
6. Cederqvist L, Garpinger O, Hägglund T, Robertsson A (2012) Cascade control of the friction stir welding process to seal canisters for spent nuclear fuel. Control Eng Pract 20(1):35–48
7. Lars Cederqvist OG, Hägglund T, Robertsson A (2010) Cascaded control of power input and welding temperature during sealing of spent nuclear fuel canisters. In: ASME 2010 dynamic systems and control conference. ASME, Cambridge, Massachusetts
8. Ross K, Sorensen C (2013) Paradigm shift in control of the spindle axis. In: Friction stir welding and processing VII, TMS (the minerals, metals & materials society), p 8
9. Wright A, Smith D, Taysom B, Hovanski Y 2021) Transitioning FSW to a controlled production process. In: Friction stir welding and processing XI, pp 91–104
10. Wright A (2021) Effective temperature control for industrial friction stir technologies. Brigham Young University, Mechanical Engineering
11. Merritt GR, Williams MB, Allison PG, Jordon JB, Rushing TW, Cousin CA (2022) Closed-loop temperature and force control of additive friction stir deposition. J Manuf Mater Process 6(5)
12. Taysom BS, Sorensen CD (2020) Adaptive relay autotuning under static and non-static disturbances with application to friction stir welding. ISA Trans 97:474–484

A Feasibility Study on Friction Screw Extrusion Additive Manufacturing of AA6060

T. C. Bor, D. H. Strik, S. Sayyad Rezaeinejad, N. G. J. Helthuis, G. S. Vos, M. Luckabauer, and R. Akkerman

Abstract Additive manufacturing in the solid state opens up possibilities for many alloys that are not suitable for fusion-based approaches. Following the advances in friction-based joining processes for high-strength aluminum and magnesium alloys, the Friction Screw Extrusion Additive Manufacturing (FSEAM) process has been developed for deposition of thin layers for cladding and additive manufacturing on a variety of substrates. In this work, the first results on the manufacturing of wall-like rectangular builds from AA6060T6 are reported. Multiple layers of about 15 mm width and 1 mm thickness were deposited with print velocities of 100–250 mm/min at constant tool rotation speed. Solid walls were formed without major macroscopic defects. Promising mechanical properties were measured with a yield strength of about 80 MPa and a tensile strength increasing from 112 to 144 MPa as function of the print velocity. The material was characterized by a fine microstructure with an average grain size below 10 μm for all builds. At the microscale, strings of unbonded regions have been observed at lower print velocities possibly related to

T. C. Bor (✉) · D. H. Strik · S. Sayyad Rezaeinejad · N. G. J. Helthuis · G. S. Vos · M. Luckabauer · R. Akkerman
Faculty of Engineering Technology, Chair of Production Technology, University of Twente, Drienerlolaan 5, 7522 NB Enschede, The Netherlands
e-mail: t.c.bor@utwente.nl

D. H. Strik
e-mail: d.h.strik@student.utwente.nl

S. Sayyad Rezaeinejad
e-mail: s.rezaei@utwente.nl

N. G. J. Helthuis
e-mail: n.g.j.helthuis@utwente.nl

G. S. Vos
e-mail: g.s.vos@utwente.nl

M. Luckabauer
e-mail: m.luckabauer@utwente.nl

R. Akkerman
e-mail: r.akkerman@utwente.nl

Y. Hovanski et al. (eds.), *Friction Stir Welding and Processing XII*, The Minerals, Metals & Materials Series,
https://doi.org/10.1007/978-3-031-22661-8_3

insufficient mixing of the deposited material with the previous layer during manufacturing leading to reduced ductility. The observed results are encouraging, indicating that additive manufacturing of aluminum alloys through FSEAM is feasible after further optimization of the process.

Keywords Additive manufacturing · AA6060T6 · Screw extrusion

Introduction

Additive manufacturing (AM) of metals has opened up a wealth of opportunities for manufacturing of special and/or unique parts and features [1–3]. Complex objects can be created layer by layer according to computer-generated designs. Function integration, efficient spare part manufacturing, and reduction of the time to market are well-known advantages.

Relevant AM approaches for metals are often based on melting and fusion of the feedstock [1–3]. Powder bed fusion, wire arc additive manufacturing, and electron beam melting all use the heat of the energy source to heat up the metal to above its melting point. Subsequently, bonding to the previously deposited layers occurs upon solidification.

The use of fusion-based AM approaches for aluminum alloys has mostly been limited to a few casting alloys, such as AlSi10Mg [4]. Important high-strength, precipitation-based alloys from the 2xxx (Al–Cu), 6xxx (Al–Si/Mg), or 7xxx (Al–Zn) series can hardly be processed or only with significant extra efforts [5, 6]. The occurring metallurgical problems originate from the inherent solid-to-liquid and liquid-to-solid phase transformations and include porosity, preferential distribution of crystal orientations (texture), solidification shrinkage, and solidification cracking.

Many of the metallurgical problems disappear if the AM process fully takes place in the solid state avoiding the liquid phase. Some approaches have been developed either based on high kinetic energy, sub-melting point sintering, or friction/plastic deformation [7]. Especially, the last approach shows huge potential as the produced parts after solid-state, friction-based AM typically contain a fine, wrought-like microstructure with small grain sizes exhibiting little texture, low residual stress levels, and virtually no porosity [8–10]. This opens up opportunities for manufacturing of severely (dynamically) loaded parts/products without costly forging routes.

In this work, the emphasis is on a new solid-state AM process that has been developed with the aim to deposit high-strength aluminum and magnesium alloys with improved microstructural properties. The Friction Screw Extrusion Additive Manufacturing (FSEAM) process is based on the merits of friction stir welding where heating of the material occurs through friction and plastic deformation. A threaded rotating tool (screw) is used to transport, extrude, and bond the softened material to the substrate. Screw dimensions can be tuned to the requested layer dimensions where the deposition width may vary from a few to tens of millimeters. The outline of the FSEAM setup is shown schematically in Fig. 1a. The rotating threaded tool is positioned just above the substrate. The material is supplied in a continuous fashion from the side through a feed channel. The feedstock may comprise wire, rods, but also powder and even chips having recycling in mind. The elevated heat and pressure densify the feedstock material and provide the necessary conditions to extrude the material through the nozzle opening. There, the rotating nature of the tool helps to spread and metallurgically bond the material to the substrate. The layer thickness depends on the distance between the tool and the substrate, (t) whereas the nominal width (W) is approximately equal to the tool tip diameter. The particular design of the setup supports almost independent control of the volumetric supply rate (V_f) and the tool rotation rate Ω.

Preliminary work conducted on the Al-Si/Mg extrusion alloy AA6060T6 has shown the importance of the mentioned process parameters. It also emphasized the role of the so-called feed ratio f which constitutes the ratio of the volume *supplied* per unit of time (V_f) over the volume *required* per unit of time ($Wt \cdot v_t$). Best results in terms of mechanical properties and the absence of volumetric defects were obtained with a feed ratio exceeding one, providing more material than strictly required.

In the current work, the role of the print/table velocity v_t was investigated where the velocity was varied between 100 and 250 mm/min while concomitantly adjusting the feed rate at a constant feed ratio. The quality of the manufactured builds was analyzed through optical and scanning electron microscopy. The mechanical properties were determined through hardness measurements and tensile tests.

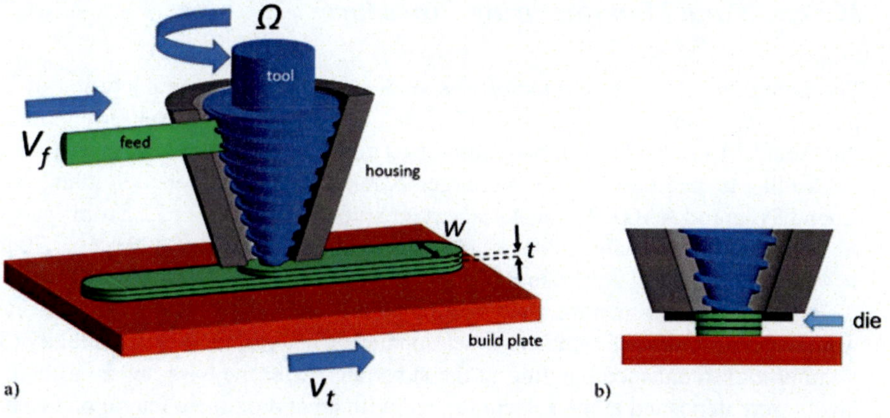

Fig. 1 **a** Fabriction screw extrusion additive manufacturing working principle. Feedstock is supplied to a rotating tool (Ω rpm). The feedstock material is softened through friction and plastic deformation and is transported to the opening at the bottom of the printhead/housing where the material is extruded and deposited on top of a substrate/build plate. The exit opening is placed at distance t above the substrate defining the layer thickness. The layer width W is determined from the volumetric supply rate V_f and the translational speed of the table v_t. **b** The printhead is equipped with a die (in black) limiting the lateral spread of the deposited layer

Experimental

Friction Screw Extrusion Additive Manufacturing Setup

The manufacturing of the builds was carried out on a modified planer machine equipped with a 13 kW electrical motor oriented vertically. A modular, in-house developed additive manufacturing system was mounted underneath. The motor drives a partly cylindrical and partly conical threaded and hardened H13 steel tool with M20 thread according to ISO965. The tool is placed centrally in a stationary housing. The normal force exerted on the AM printhead in the direction parallel to the axis of the tool was measured with three load cells placed equally around the circumference in the motor frame to monitor the progress of the AM process. Several K-type thermocouples were inserted close to the 10 mm nozzle opening to record the print temperature development over time. The nozzle opening was also equipped with a small die comprising support walls on either side of the nozzle opening to better control the shape and dimensions of the deposited layer during the deposition process, see Fig. 1b. A cooling system was integrated into the housing around the screw and the feeding system. A hydraulic system was used to supply feed material in a controlled fashion. Measurement data were stored employing a National Instruments data acquisition system.

Materials and Manufacturing Procedure

The influence of the print velocity was studied through the manufacturing of four builds with a print velocity of 100, 150, 200, and 250 mm/min at 400 rpm, a layer thickness of 1 mm and a nominal width of 10 mm. The feed ratio was approximately 1.3 leading to an actual layer width larger than the above-mentioned nominal value. Round 8 mm rod AA6060T6 material cut into pieces of 45 mm or 22.5 mm was used as feed material. Available plates of AA2024 with dimensions of 300 mm × 70 mm × 4 mm were used as substrate material.

Before the actual manufacturing of the build started, four, so-called, base layers were deposited with the same cross-sectional dimensions but at lower velocity (35–50 mm/min) to enhance bonding to the substrate. After the base layers in total, 50 layers were deposited at the selected speed with an approximate length of 140 mm. The feed system was refilled with the required amount of feed material after each deposited layer.

Mechanical and Microstructural Analysis

After manufacturing of the build and a waiting time of least two weeks to allow for post-processing microstructural changes, the builds were cut into pieces, where one piece was used for analysis of build quality, microstructure and hardness, and another piece was used to extract tensile test specimens. The approximate locations of both pieces are shown in Fig. 2. The first piece was prepared following standard metallurgical procedures. From the second piece, tensile test specimens were extracted employing electric discharge machining. The 45 mm long and 2 mm thick test specimens exhibited a 20 mm gauge section (cross-sectional dimensions 5 mm × 2 mm) and had a smooth transition from the clamping section (dimensions 10 mm × 2 mm) with a 2.5 mm filet radius. For each build, three tensile test specimens through the thickness were obtained numbered 1–3 with number 2 located in the build midplane in the width direction. Tensile tests were performed with a Zwick 100 SMZ100/TL3A equipped with a GTM 100 kN force cell and a 066608 Multisens extensometer at 1 mm/min crosshead displacement.

A Keyence VHX 7000 light microscope and a JEOL JSM 7200F Scanning Electron Microscope (SEM) were used for microstructural analysis. A Leco LM 100AT hardness tester equipped with a micro-Vickers hardness indenter was used to assess the hardness distribution employing 300 gf and a waiting time of 15 (s).

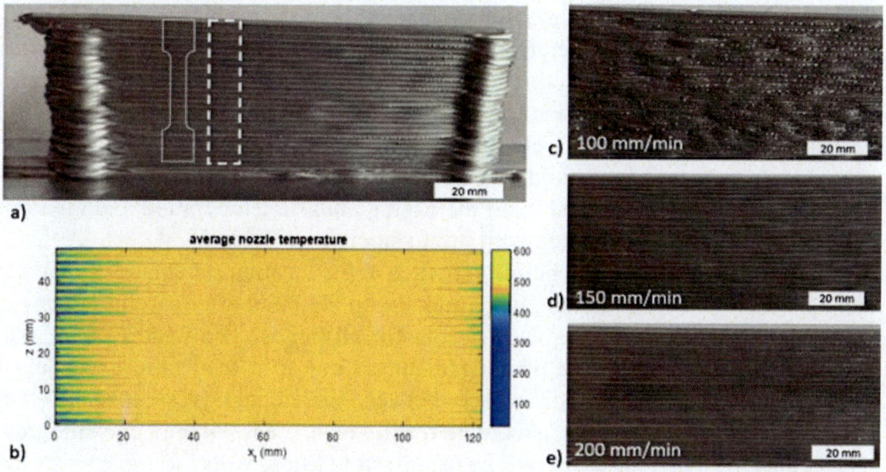

Fig. 2 **a** AA6060 build manufactured at AA2024 substrate at table velocity of 250 mm/min. **b** Spatial distribution of measured nozzle temperature during manufacturing at 250 mm/min. **c–e** Centre regions of builds manufactured at 100 mm/min, 150 mm/min and 200 mm/min, respectively. Individual layers are clearly visible. Locations of tensile test specimen extractions (3 through the thickness) and hardness/microscopy are indicated in (**a**) by dotted and dashed lines, respectively. Locations hold for all builds

Table 1 Average nozzle temperature <T_{nozzle}> and exerted normal force <F_n> during FSEAM of AA6060T6 for a range of print velocities v_t. Average values of yield strength <σ_{yield}> and tensile strength <$\sigma_{tesnile}$> , and hardness <HV> as obtained from tensile tests and Vickers microhardness measurements, respectively

v_t (mm/min)	100	150	200	250
<T_{nozzle}> (°C)	384 ± 35	485 ± 21	493 ± 28	525 ± 10
<F_n> (kN)	5.3 ± 1.4	8.8 ± 1.6	7.3 ± 1.0	9.3 ± 1.6
<σ_{yield}> (MPa)	80.8 ± 1.7	75.7 ± 2.2	79.6 ± 3.5	84.9 ± 3.9
<$\sigma_{tensile}$> (MPa)	112.4 ± 8.8	114.0 ± 19.6	124.4 ± 14.9	144.5 ± 0.2
<HV>	39.5 ± 1.9	39.1 ± 1.5	39.7 ± 2.8	41.6 ± 1.8

Results and Discussion

Build Appearance and Conditions

The FSEAM experiments were carried out successfully, and four AA6060 builds were manufactured on top of an AA2024 substrate. The build produced at 250 mm/min is shown in Fig. 2a. The deposited layers are clearly visible. The layer widths at the build ends are somewhat larger due to the relatively low velocity employed during vertical motion of the printhead for the volumetric supply rate employed. The temperature measured close to the nozzle opening while printing is displayed in Fig. 2b. Temperatures were fairly constant during the manufacturing process except for the respective ends on either side of the build where the temperature temporarily dropped due to the refeeding process, see also Sect. 2.2. Builds manufactured at lower table velocities looked similar, see Fig. 2c–e for a display of the middle sections of the builds. The build deposited at 100 mm/min showed some signs of waviness. In general, both the average nozzle temperature and the normal force show an increasing trend with print velocity, see Table 1.

The cross sections of the builds extracted at the location schematically indicated in Fig. 2 show the individual layers making up the build after careful etching, see Fig. 3. Macroscopic defects were absent for all builds. The widths of the builds were all beyond the nozzle opening diameter due to a feed ratio exceeding the value of one. The gray tone differences between individual layers, especially visible for the 200 mm/min sample, cannot be readily explained. Possible crystallographic variations from layer to layer will be examined in future work.

Mechanical Testing

The hardness values were determined in the midplane of each build on the cross sections shown in Fig. 3. The measurement results are shown in this figure on the right, and average values are collected in Table 1. The hardness remains more or less

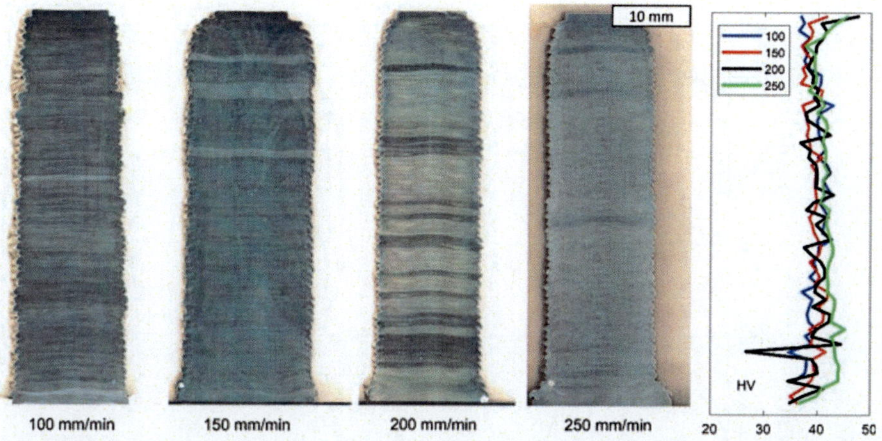

Fig. 3 Cross sections of manufactured builds at 100, 150, 200, and 250 mm/min. Individually deposited layers are visible. Measured Vickers hardness values determined in the midplane along the build direction, are shown on the right. The scale bar applies to all builds

constant through the height of the build and without significant differences among the various builds; i.e., the hardness does not seem to be dependent on the printing velocity employed, despite the differences in average nozzle temperature (see Table 1). The hardness is reduced with respect to the hardness of the feed material of 80 HV which is ascribed to the thermomechanical nature of the process leading to changes in the size and distribution of the strengthening precipitates.

The results of the tensile tests are displayed in Fig. 4a–d. In all cases, plastic deformation of the test specimen occurred. However, for some samples, failure occurred prematurely. The test specimens of the 100 mm/min build broke relatively early in the plastic region leading to limited elongation at fracture. These samples all broke in the filet region of the test specimens, i.e., outside of the gauge section. Similar behavior was also observed with 150.1 and 150.2. Samples 150.3, 200.1, and 250.1 showed failure close to one of the attachment points of the strain gauge explaining the deviating stress–strain behavior. In case of the 250 mm/min sample, significant plastic deformation to well beyond 0.3 (30%) occurred for samples 250.2 and 250.3. The average results per build were collected in Table 1. The values of the yield strength were approximately equal at values around 80 MPa, but the tensile strength seemed to increase with higher printing velocities from 112 MPa at 100 mm/min to 145 MPa at 250 mm/min. The elongation at fracture could only be determined accurately for those samples fractured within the spanned region of the strain gauge. Values increase from approximately just below 20% for the 150 mm/min build to above 40% for the 250 mm/min build. The mechanical properties derived from the 250 mm/min build are comparable to those achieved through Additive Friction Stir–Deposition (AFS–D) of AA6061 by Rutherford et al. [11].

The fracture surfaces of the tensile test specimens were analyzed through light microscopy and scanning electron microscopy. The results of two samples, one with

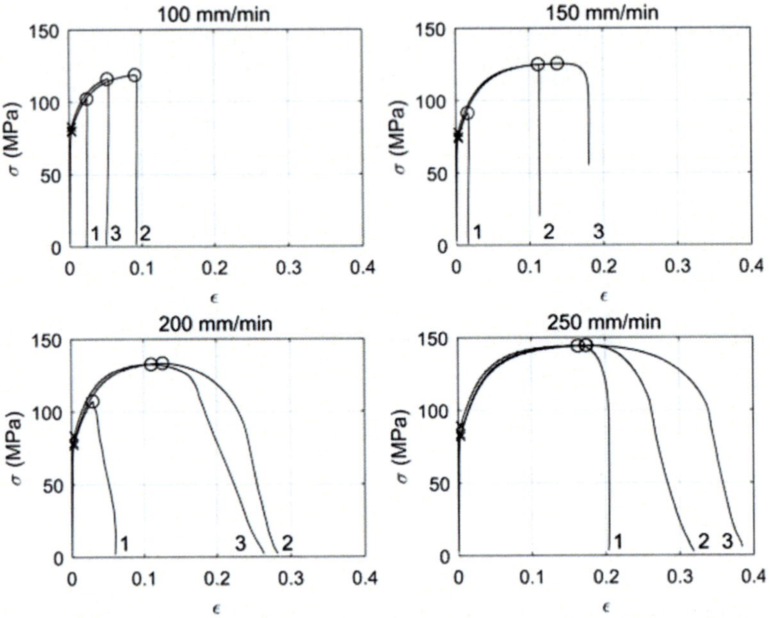

Fig. 4 Tensile test results of test samples of builds manufactured at 100, 150, 200, and 250 mm/min. All samples were extracted in the build direction (see Fig. 2) and numbered 1–3, where sample 2 is positioned in the midplane. The 0.2% yield strength and the tensile strength for each sample are given by the "x" and "o" symbols, respectively. A sudden decrease of the stress beyond the tensile strength indicates a necking event outside of the strain gauge section (see Sect. 2.2). Average results are displayed in Table 1

high elongation at fracture (250.3) and the other with a low value (100.1), are shown in the SEM images in Fig. 5. The fracture surface of the 250.3 sample visible in Fig. 5a reflects the ductile nature of the sample with elongation at fracture beyond 30% (see Fig. 3) displaying dimples over a large portion of the fracture surface as clearly visible in the magnified rectangular region shown in Fig. 5b. The 100.1 sample also shows signs of significant plastic deformation through the presence of dimple-rich regions but also shear-like regions without clear signs of plastic deformation are visible adjacently explaining the much lower values of the elongation at fracture of this sample. The fracture surfaces of the other samples showed similar patterns with a greater portion of dimple-rich regions as the elongation at fracture increased. Hence, local deficiencies in bond quality, as recognized by lack of dimple-rich regions, may explain premature tensile test failure.

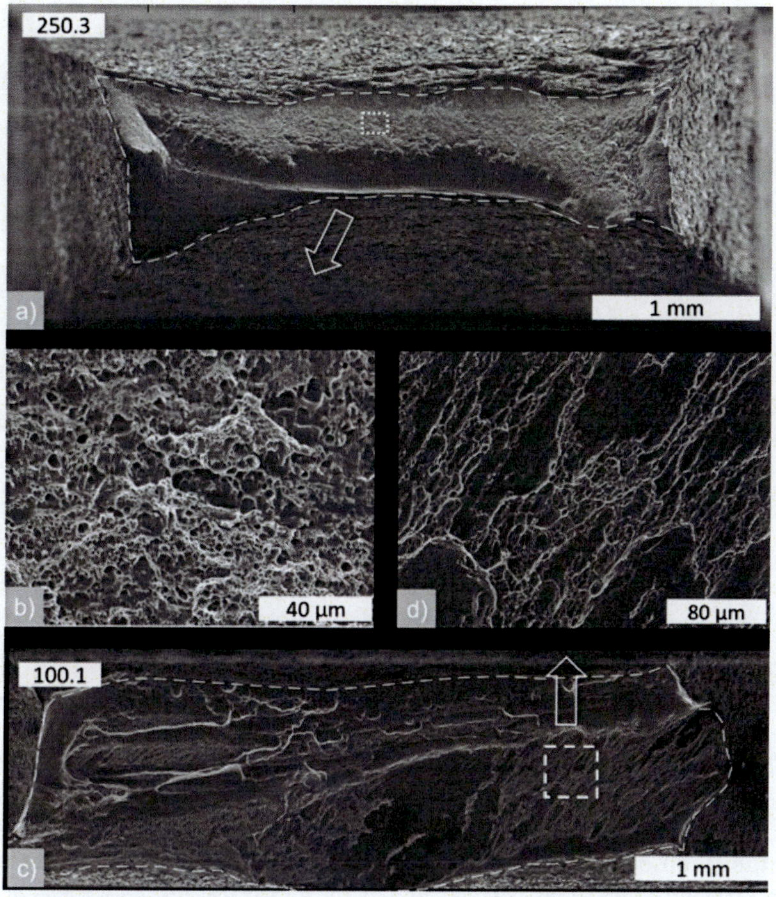

Fig. 5 **a** Fracture surface of sample 250.3 after tensile test. The dashed line indicates the facture surface contour. Magnified detail of fracture surface shown in **b** displaying dimples in line with >30% elongation at fracture. **c** Fracture surface of sample 100.1 after tensile test. Magnified detail shown in **d** displays mixed mode fracture with both dimple-rich regions and shear-like fracture in line with limited ductility observed in tensile test

Microscale Analysis

The cross sections displayed in Fig. 3 were analyzed in greater detail before etching and hardness analysis were done. At higher magnifications, the presence of manufacturing defects became apparent for the 100 mm/min build as visible in Fig. 6a. They constitute a string of small, crack-like, unbonded regions oriented approximately horizontally. In some occasions, the series of defects was oriented more upwards or downwards. It seems likely that, given the nature of the AM process, these unbonded regions formed at the interface when the newly deposited layer was brought into contact with the underlying substrate. The defects were present throughout the entire

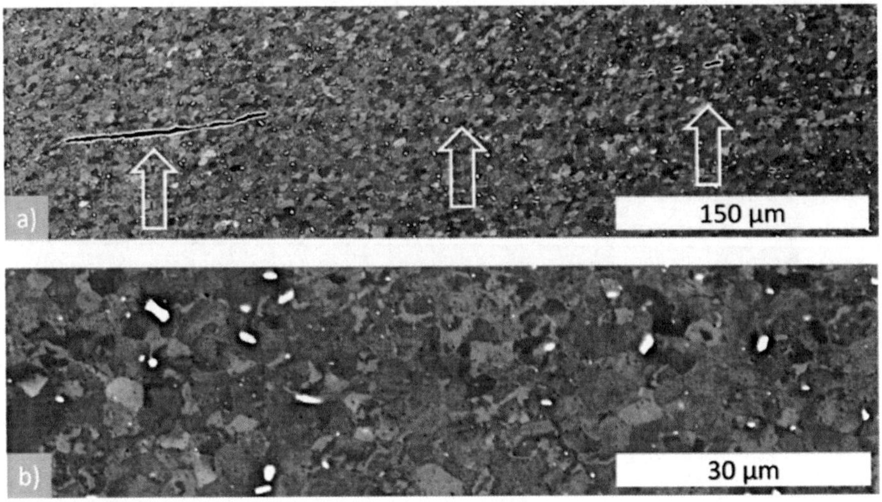

Fig. 6 Cross section of sample 100.1 at higher magnification: **a** a string of crack-like unbonded regions present as indicated by arrows. **b** Observed microstructure with relatively small grains typically smaller than 10 μm (in various tones of gray) and a number of precipitates (in white)

cross section of the analyzed sample. They were also observed in the cross sections of other builds but to a lesser extent. In the builds manufactured at 250 mm/min, they were hardly observed and rather small. A series of such small and aligned defects make the build vulnerable to sudden crack growth upon tensile loading during a tensile test accelerating premature failure. In this way, the ductility development (elongation at fracture) of the various builds could be explained, where the lowest ductility is observed for the samples containing the highest presence of these defects (e.g., 100 mm/min). It also explained why the tensile test results were so dependent on the print velocity, whereas the hardness values were not. A hardness tests relies mostly on plastic deformation under compression, where the presence of crack-like defects is hardly relevant.

At higher magnifications also, the microstructure became apparent. The grain size was typically smaller than 10 μm for all builds and also did not vary much through the height and width of the sample, see Fig. 5b for a micrograph of the 100 mm/min build. A more in-depth analysis employing SEM–electron backscatter diffraction (EBSD) is to be performed at a later stage which will also help in understanding the location of the strings of defects observed in relation to the deposited layers (inside a deposited layer vs at the interface of adjoining layers) as the typical crystalline orientation for consecutively deposited layers often varies due to a difference in deposition direction (forward vs. backward deposition).

Bonding Model Description

Additive manufacturing relies on the bonding of consecutive layers one after another. Friction and plastic deformation play an important role in bonding of layers in the solid state. The added material needs to be brought in sufficiently intimate contact with the substrate. Moreover, the oxide layers present at the interfaces need to be broken up to allow metal atoms to come into close enough contact for metallic bonding to occur. For example, in FSW, it is the tool pin, while aligned with the mating interfaces, that fulfills this role.

In the present setup, the bottom of the screw is placed at some distance from the substrate without actual contact with the substrate (see Fig. 1). The substrate oxide layer can only be broken up indirectly by shear and normal forces transferred through the supplied material. It explains the use of a feed ratio larger than one. In this way, considerable plastic deformation of the layer(s) underneath the printhead must occur to accommodate the excess volume supplied. However, it seems that the current manufacturing conditions, given by the feed ratio, tool rotation rate, layer dimensions and printing velocity, may still not be optimal for all builds. From the stacked appearance of the layers in Fig. 3, it may be concluded that no significant mixing of deposited layers occurred, at least not at macro-scale, despite the plastic deformation imposed. However, also at microscale mixing was probably insufficient seeing the appearance of strings of unbonded, crack-like defects (Fig. 6a), and the fracture surfaces where regions with and without dimples were located next to each other (Fig. 5a). Current work is directed toward further understanding of the role of the process conditions and tool design in the deposition and bonding process.

Conclusions and Outlook

In this work, the first results on the manufacturing of wall-like rectangular builds from AA6060T6 were reported employing friction screw extrusion additive manufacturing. Multiple layers of about 15 mm width and 1 mm thickness were deposited with a print velocity of 100 mm/min to 250 mm/min. Tool rotation speed was held constant at 400 rpm, and the volumetric flow rate of the supplied material was adjusted to match the print velocity. Solid walls were formed without major macroscopic defects. The material was characterized by a fine microstructure with an average grain size below 10 μm for all builds irrespective of the print velocity.

The mechanical properties of the manufactured builds showed promising results. The yield strength was about 80 MPa for all builds, and the tensile strength increased from just above 112 MPa at 100 mm/min to 145 MPa at 250 mm/min. The average hardness was measured at approximately 40 HV independent of print velocity.

The fracture surfaces of the tensile test samples displayed a mostly ductile appearance for the samples manufactured at 250 mm/min. At lower print velocities, the fracture surfaces showed a mixed appearance with ductile regions next to regions displaying less ductile, shear-like fracture behavior. Elongations at fracture were correspondingly lower.

Microscale analysis of the cross sections of the builds revealed the presence of small manufacturing defects comprising strings of unbonded, crack-like regions, especially for the builds made at lower velocity explaining further the limited ductility of the tensile tests. Insufficient mixing of deposited material with the substrate at microscale could be a possible origin.

The observed results are encouraging indicating that additive manufacturing of aluminum alloys through FSEAM is feasible after further optimization of the process.

Acknowledgements The authors acknowledge the financial support of the Dutch Research Council (OCENW.XS21.1.078). We also thank Norbert Spikker and Kai Wissink of Geerdink for their specialist help with the manufacturing setup.

References

1. Frazier WE (2014) Metal additive manufacturing: a review. J Mater Eng Perf 23:1917–1928
2. Blakey-Milner B, Gradl P, Snedden G, Brooks M, Pitot J, Lopez E, Leary M, Berto F, du Plessis A (2021) Metal additive manufacturing in aerospace: a review. Mater Des 209:110008
3. DebRoy T, Wei HL, Zuback JS, Mukherjee T, Elmer JW, Milewski JO, Beese AM, Wilson-Heid A, De A, Zhang W (2018) Additive manufacturing of metallic components—process, structure and properties. Progress Mater Sci 92:112–224
4. Zhang D, Sun S, Qiu D, Gibson MA, Dargusch MS, Brandt M, Qian M, Easton M (2018) Metal alloys for fusion-based additive manufacturing. Adv Eng Mater 20:1700952
5. Mukherjee T, Zuback JS, De A, DebRoy T (2016) Printability of alloys for additive manufacturing. Sci Rep 6:19717
6. Martin J, Yahata BD, Hundley JM, Mayer JA, Schaedler TA, Pollock TM (2017) 3D printing of high-strength aluminium alloys. Nature 549:365–369
7. Yu HZ, Mishra RS (2021) Additive friction stir deposition: a deformation processing route to metal additive manufacturing. Mater Res Lett 9(2):71–83
8. Yu HZ, Jones ME, Brady GW, Griffiths RJ, Garcia D, Rauch HA, Cox CD, Hardwick N (2018) Non-beam-based metal additive manufacturing enabled by additive friction stir deposition. Scripta Mat 153:122–130
9. Blindheim J, Gronga Ø, Aakenes UR, Welo T, Steinert M (2018) Hybrid Metal Extrusion & Bonding (HYB)—a new technology for solid-state additive manufacturing of aluminium components Proc. Manufac 26:782–789
10. Xie R, Shi Y, Liu H, Chen S (2021) A novel friction and rolling based solid-state additive manufacturing method: microstructure and mechanical properties evaluation. Mater Today Comm 29:103005
11. Rutherford BA, Avery DZ1, Phillips BJ, Rao HM, Doherty KJ, Allison PG, Brewer, LN, Jordon BJ (2020) Effect of thermomechanical processing on fatigue behavior in solid-state additive manufacturing of Al-Mg-Si alloy. Metals 10:947

Part II
Friction Stir Processing

Enhanced Tensile and Tear Toughness Properties of Thin-Wall Vacuum-Assisted High-Pressure Die-Cast Aural-5 Alloy by Friction Stir Processing

Avik Samanta, Hrishikesh Das, Glenn J. Grant, and Saumyadeep Jana

Abstract Steadily rising demand for glider weight reduction has driven the development of vacuum-assisted high-pressure die-cast (HPDC) Al alloys for automotive structural components. Aural-5 is a strontium-modified HPDC alloy utilizing manganese (Mn) to reduce die soldering, eliminating detrimental needle-shaped Fe-bearing β-phase intermetallic and improving ductility. HPDC Aural-5 contains shrinkage porosity, dendrites with Al-Si eutectic colonies, externally solidified crystals (ESCs), shear/band structure, and large second-phase particulates. Porosity, ESCs, and large second-phase particles work as crack initiation sites, negatively impacting tensile properties. In this study, friction stir processing (FSP) is employed for microstructural modification of a thin-walled HPDC Aural-5 by eliminating porosity and breaking down dendrites, second-phase particles, eutectic colonies, ESCs, and shear/band structures to create wrought microstructure with homogenized particle distribution. Mechanical property characterization indicates ~30 and ~35% enhancement in yield strength and ductility and associated marked effects on ~69% improvement in tear toughness according to the ASTM B871 test.

Keywords Friction stir processing · Aural-5 · Strength · Tear toughness · High pressure die-casting

Introduction

In recent decades, automotive original equipment manufacturers have been trying to produce a low-cost solution for vehicle lightweighting to meet zero carbon emission goals. To achieve that target, high-pressure die-cast (HPDC) aluminum castings alloys are increasingly used/considered for structural applications, e.g., shock towers, pillars, and floor rails for weight reduction. Multi-component, heavy structural steel load-bearing structural assemblies were replaced with single HPDC Al casting with intricate structural profiles to achieve similar functionality with less

A. Samanta · H. Das · G. J. Grant · S. Jana (✉)
Pacific Northwest National Laboratory, Richland, WA 99352, USA
e-mail: Saumyadeep.Jana@pnnl.gov

Y. Hovanski et al. (eds.), *Friction Stir Welding and Processing XII*,
The Minerals, Metals & Materials Series,
https://doi.org/10.1007/978-3-031-22661-8_4

unpredictability [1]. The use of HPDC Al casting provides the opportunity for a reduction in part counts and weight. With the advancement of electric vehicles, hybrid powertrains and vehicle electrification are becoming ampler in the automotive industry. For lightweighting electric vehicles, HPDC Al castings can play a crucial role in designing lightweight housing for batteries and various powertrain and transmission components. Typical HPDC Al alloys use high Fe in alloy composition to avoid die soldering. However, high Fe content leads to the formation of needle-shaped β-$AlFeSi_5$ phases, which has harmful effects on the overall mechanical properties, especially ductility. Moreover, HPDC Al alloys contain gas and shrinkage porosity which act as a stress concentration and crack initiation sites under mechanical loading. HPDC Al alloys also contain acicular Si and large second-phase particulates, which negatively impact tensile strength and ductility [2, 3]. Therefore, the combination of a number of microstructural features and the alloy chemistry negatively impacts the mechanical properties [4] of HPDC Al alloys.

Improvements in die-casting technology (vacuum-assisted HPDC method and high vacuum die-casting method) help in reducing casting defects such as porosity [5]. Alloy chemistry is often altered to improve mechanical properties. Several low Fe containing premium HPDC Al alloy compositions (Silafont, Castasil, Aural, etc.) were developed with improved mechanical properties [6]. Adding 50–200 ppm of strontium (Sr) can transform acicular Si into a fine, fibrous structure [7, 8]. The ductility and manufacturability are significantly improved with the addition of Sr modifier without compromising strength. It satisfies the growing interests of the HPDC aluminum market. Although vacuum-assisted HPDC process with Sr- and more Mn-based material chemistry helped some of the issues of HPDC Al alloys, they still consist of shrinkage porosity in the middle section, dendritic microstructure with Al-Si eutectic colonies, shear/band structure beneath the die-wall, and large second-phase particulates.

This study applies friction stir processing (FSP) on thin-walled vacuum-assisted HPDC Aural-5 alloy, a popular Sr-modified premium quality Al alloy primarily used for automobile structural components. FSP is a widely used thermomechanical processing tool to effectively eliminate casting defects and modify the cast microstructure of aluminum alloys [9, 10]. FSP imposes severe plastic deformation on the workpiece material through a rotating tool-driven complex material mixing and microstructural change. As a result, the mechanical and metallurgical properties of workpiece material are altered through dynamic recrystallization and refinement of constituent phases [11]. FSP can elevate the tensile properties of HPDC Al alloys through porosity elimination and microstructure refinement [12]. FSP-driven microstructure modification can lead to enhanced strength and ductility [13] and associated marked effects on improved resistance to fatigue failure [14] and fracture toughness [15]. This study investigates FSP-driven microstructural modification on Aural-5 and change in tensile properties. Finally, tear toughness and strength are compared between FSP and without FSP according to the ASTM B871 test.

Experimental Methods

This study conducted FSP trials on HPDC Aural-5, a Sr-modified and low Fe (~0.09 wt.%) containing alloy. FSP experiments were carried out on flat plate geometry. The thickness of Aural-5 plates was measured to be 2.4 ± 0.1 mm. The nominal composition of Aural-5 is reported in Table 1.

For FSP experiments, a scrolled shoulder tool was used, as shown in Fig. 1a. The tool has a conical threaded pin with three flat segments equally separating the threads. Position control mode is used during FSP experiments. Penetration depth was kept at 2.15 mm. Tool rotation is counterclockwise at 2° tilt angle for single-pass FSP experiments. Therefore, the advancing side (AS) and retreating side (AS) are illustrated in Fig. 1b. After several initial experiments, one FSP condition was chosen, which gives defect-free microstructure consolidation. The tool rotational speed was kept at 1200 RPM, and the traverse speed was 0.7 m/min.

Metallographic specimens were extracted from the as-cast and nugget zone of FSPed samples for microstructure characterization. In addition, full-thickness subsize ASTM E8 specimens were machined from the middle of the nugget region, as shown in Fig. 1b, and from an unprocessed Aural-5 plate to characterize the tensile properties. Gauge length and width of bulk tensile specimens were kept at 15.8 and 3.4 mm. During the tensile testing, a computer-controlled tester set the initial crosshead velocity as 0.005 mm s^{-1} at room temperature.

Table 1 Elemental composition of Aural-5

Al	Si	Mg	Mn	Fe	Ti	Sr
Balance	6.88	0.25	0.56	0.09	0.06	125 ppm

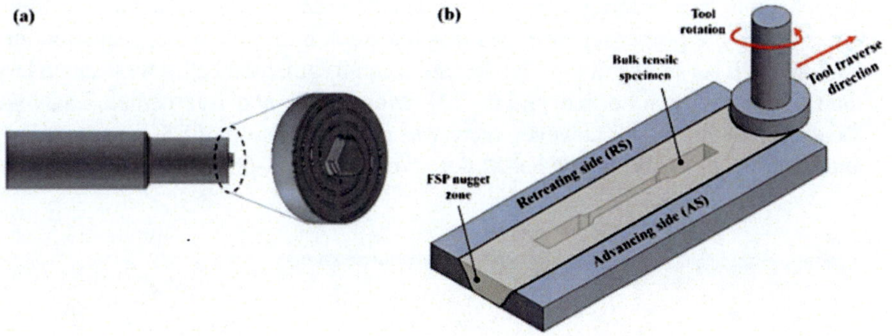

Fig. 1 **a** Schematic representation of the FSP tool used in the study, **b** location of tensile specimen with respect to FSP nugget

Results and Discussion

Microstructure Evolution

The microstructure of as-cast HPDC Aural-5 is characterized before performing any FSP experiments. A typical low-magnification full-thickness cross-sectional image of HPDC Aural-5 is shown in Fig. 2a. Most of the porosity is concentrated near the middle section, whereas the die-wall areas are mostly porosity-free. As illustrated in Fig. 2b, which is a higher magnification view of the red rectangle of Fig. 2a, shows three distinct layers of microstructure: (1) a surface layer or die-wall, (2) a middle layer or mid-wall, and (3) a layer between the surface layer and middle layer (the so-called shear/defect band). The surface layer is ~200–250 μm thick with a very refined microstructure and little to no porosity. The middle layer contains high porosity. Most porosity is shrinkage porosity, as illustrated in Fig. 2c. The die-casting industry typically observes these two layers [16]. Less porous, refined microstructure at die-wall is generated through faster solidification during the HPDC process, and a relatively slower solidification rate determines coarse porous microstructure of mid-wall. A third layer of microstructure named as shear/defect band is observed for Aural-5 in between die-wall and mid-wall. A shear band is typically formed in some HPDC and vacuum-assisted HPDC Al and Mg alloy containing very fine and dendritic primary grains with high percentage of eutectic colonies [17]. It is formed by slip between the surface layer and the inner adjacent region due to significant shear occurring either during die filling or during pressure intensification [18].

A typical higher magnification micrograph of HPDC Aural-5 is shown in Fig. 3a, indicating that it contained large dendritic externally solidified crystals (ESCs) and disintegrated ESCs in between the in-cavity solidified primary grains and eutectic (marked by blue arrows). The ESCs are nucleated and grown externally within the shot chamber before being injected into the HPDC cavity. ESCs can have an adverse effect on tensile properties by expediting the crack initiation and propagation under loading [19]. As shown in Fig. 3a, the ESCs and fragmented ESCs were much larger than the in-cavity solidified grains [17]. Most ESCs and fragmented ESCs were located in the mid-wall; however, there was a small fraction of ESCs in the die-wall and shear band. Additionally, Fe–Mn-containing second-phase particles can be seen

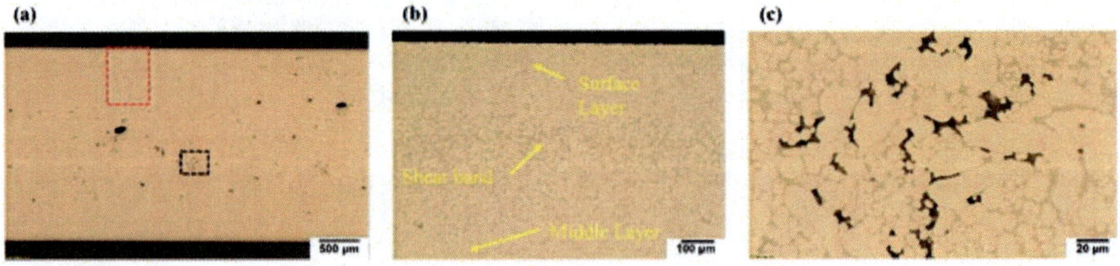

Fig. 2 Microstructure of vacuum-assisted HPDC Aural-5 plate: **a** overview of the cross-section; **b** shear/defect band formation below die-wall/skin; and **c** shrinkage porosity in the middle section

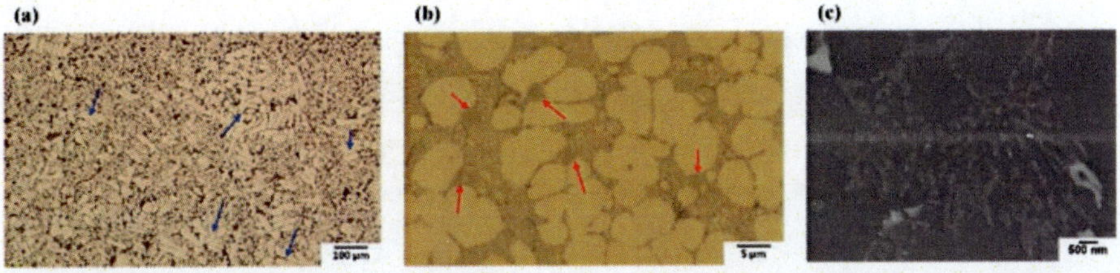

Fig. 3 Presence of **a** a mixture of large dendritic ESCs and fragmented ESCs; **b** large second-phase particles; and **c** fine fibrous eutectic Si structure

between the dendrites (marked by red arrows in Fig. 3b). The addition of Sr led to an extensive refinement of the morphology of eutectic Si. As illustrated in Fig. 3c, Al–Si eutectic colonies contain fine fibrous networks of eutectic Si.

To study cross-sectional micrographs of FSPed plate, metallography specimens were extracted at 125 mm away from the plunge location. Figure 4a illustrates a microstructural overview at low-magnification of Aural-5 after FSP. The basin-shaped nugget region has distinct AS and more diluted RS boundaries. Two locations within the nugget zone of the FSP2 were chosen for a detailed study of the microstructure. In Fig. 4b, a high magnification image of the FSP-HPDC interface shows the transition of the FSPed microstructure to the parent HPDC microstructure. A gradual change is observed from the dendritic structure in the as-cast region to the wrought microstructure in the processed zone. Within the nugget zone, as-cast dendrites are disintegrated to form a wrought microstructure. A thermomechanical affected zone (TMAZ) works as a transition zone from the severe plastic deformation zone to the as-cast material. In Fig. 4c, a high magnification image of the nugget zone shows that dendrites and Al-Si eutectic colonies were fragmented. This location was at the same position as the surface-layer edge. Therefore, it can be concluded that the FSP processing eliminated the high-density eutectic regions and made a more uniform distribution of elements across the cross section. In addition, the ESCs were also eliminated after FSP. Furthermore, the shrinkage porosities were removed completely from the middle section after FSP, and second-phase particles were refined.

Tensile Properties

Eight full thickness E8 specimens for HPDC condition and three for FSP condition were tested. Figure 5 compares the engineering stress versus percent elongation plot between two specimens from HPDC and FSPed conditions. Those two specimens denoted the maximum and minimum ductility of each condition, and others fell in between. It shows that FSP can increase yield strength and % elongation to failure after processing. The yield strength (YS) of HPDC Aural-5 alloy is ~101.3 ± 5.5 MPa, whereas that is increased by ~30% to ~131.5 ± 3.8 MPa after FSP. The change in

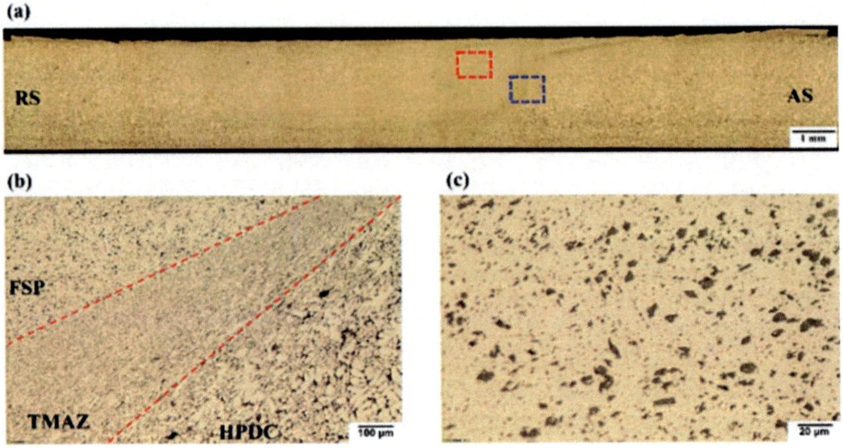

Fig. 4 Cross-sectional microstructural features of FSPed Aural-5 alloy plate: **a** nugget zone shape and size; **b** interface between HPDC and FSP regions; **c** wrought microstructure with uniformly distributed Si and second-phase particles

UTS is statistically insignificant. It is ~243.8 ± 6.6 MPa after FSP compared to 257.4 ± 11.7 MPa for the as-cast Aural-5 condition. Aural-5 was a highly ductile material; however, FSP was still able to improve the ductility further. The elongation to failure for HPDC Aural-5 was 13.5 ± 3.8%, whereas after FSP, it increased by 35% to 18.3 ± 1.7%.

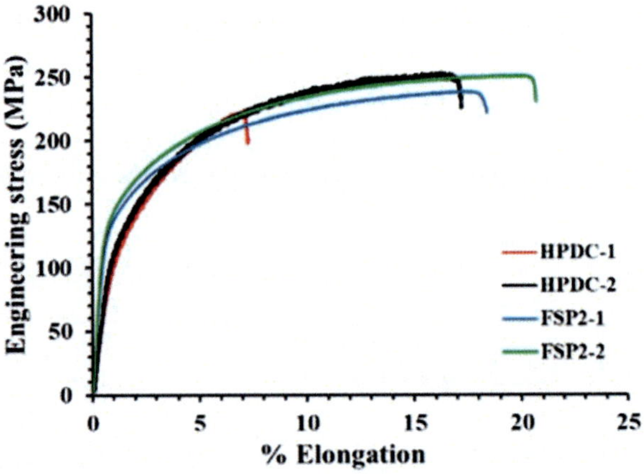

Fig. 5 Comparison of engineering stress versus percent elongation between HPDC and FSP condition

Tear Toughness

According to ASTM standard B871-01 [20], standard tear test specimens were machined for HPDC and FSPed material. The schematic of specimen geometry is illustrated in Fig. 6a. The rectangular-shaped tear test specimens have an 11 mm deep 60° V-notch with a 0.025 mm notch radius. For FSP, the specimens were machined in two different orientations. In transverse orientation, the notch tip was placed 2 mm inside the boundary of the AS. The notch tip was placed along the centerline of the tool traverse direction for longitudinal orientation. After placing the specimen in MTS 810 materials testing system with a 10 kN load cell, 220 N preload was applied. During the test, a loading rate of 0.02 mm/s was used at ambient temperature and force–displacement data were recorded. In addition, thickness and net section width between the notch root and back edge of the specimen was measured, and the load–displacement data were normalized accordingly.

The tear test is very useful as an indicator of the toughness of thin aluminum sheets. Representative force–displacement curves of HPDC and two orientations of FSPed Aural-5 are illustrated in Fig. 6b. The slopes of the loading part of the curves are very similar to each other. However, both FSP configurations required higher force to initiate crack than HPDC. Additionally, FSP in longitudinal configuration needed more force than FSP in transverse configuration. Total energy required for crack initiation and propagation is calculated by numerical integration of the area under the force–displacement curve. As illustrated in Fig. 6c, FSPed specimens required higher energies for crack initiation and propagation than HPDC specimens. Therefore, FSP treatment on Aural-5 induced higher tear toughness and resistance to crack initiation and propagation than the HPDC condition in longitudinal and transverse orientation.

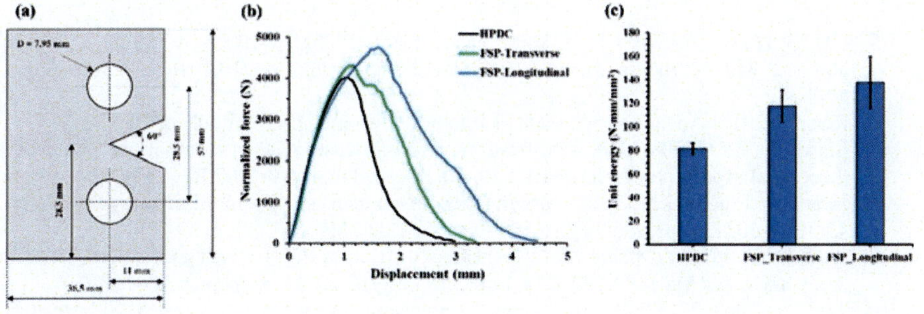

Fig. 6 Tear toughness test: **a** sample geometry; and comparison of **b** the force–displacement curve and **c** unit total energy

Conclusion

The microstructure of vacuum-assisted HPDC Aural-5 alloy consists of process-related high shrinkage porosity in the mid-wall area, α-Al dendrites, presence of ESCs, surface-layer edge beneath the die-wall with a high volume fraction of Al–Si eutectic colonies and large second-phase particles. Even though Sr modification significantly refined eutectic silicon and vacuum assistance reduced porosity in Aural-5, there is significant variation in the strength and ductility. After FSP, the as-cast microstructure is substantially modified by eliminating porosity, breaking down α-Al dendrites into a wrought microstructure. It also eliminates ESCs and breaks down Al-Si eutectic colonies into uniform distribution of fragmented Si particles. The shear/band structure is also broken down during FSP. The Fe–Mn-containing second-phase particles are also refined to some extent and redistributed homogeneously across the processed zone. This microstructure modification significantly improved tensile strength and ductility. Additionally, FSP-modified microstructure demonstrates more resistance to tear and an increase in tear toughness.

References

1. COSMO International (2014) Aluminum high pressure die casting
2. Hartlieb M (2013) Aluminum alloys for structural die casting. In: Die casting engineer, pp 40–43
3. Outmani I, Fouilland-Paille L, Isselin J, el Mansori M (2017) Effect of Si, Cu and processing parameters on Al-Si-Cu HPDC castings. J Mater Process Technol 249:559–569. https://doi.org/10.1016/j.jmatprotec.2017.06.043
4. Sigworth GK, Donahue RJ (2021) The metallurgy of aluminum alloys for structural high-pressure die castings. Int J Metalcast 15:1031–1046. https://doi.org/10.1007/s40962-020-00535-x
5. Niu XP, Hu BH, Pinwill I, Li H (2000) Vacuum assisted high pressure die casting of aluminium alloys. J Mater Process Technol 105:119–127. https://doi.org/10.1016/S0924-0136(00)00545-8
6. Rowe J (2012) Advanced materials in automotive engineering. Woodhead Publishing
7. Sigworth GK (2008) The modification of Al-Si casting alloys: important practical and theoretical aspects. Int J Metalcast 2:19–40. https://doi.org/10.1007/BF03355425
8. Hegde S, Prabhu KN (2008) Modification of eutectic silicon in Al-Si alloys. J Mater Sci 43:3009–3027. https://doi.org/10.1007/s10853-008-2505-5
9. Sharma SR, Ma ZY, Mishra RS (2004) Effect of friction stir processing on fatigue behavior of A356 alloy. Scr Mater 51:237–241. https://doi.org/10.1016/j.scriptamat.2004.04.014
10. Tsai FY, Kao PW (2012) Improvement of mechanical properties of a cast Al-Si base alloy by friction stir processing. Mater Lett 80:40–42. https://doi.org/10.1016/j.matlet.2012.04.073
11. Sun N, Apelian D (2011) Friction stir processing of aluminum cast alloys for high performance applications. JOM 63:44–50. https://doi.org/10.1007/s11837-011-0190-3
12. Samanta A, Seffens RJ, Das H et al (2022) Microstructure-refinement–driven enhanced tensile properties of high-pressure die-cast A380 alloy through friction stir processing. J Manuf Process 78:352–362. https://doi.org/10.1016/j.jmapro.2022.04.027
13. Rao AG, Deshmukh VP, Prabhu N, Kashyap BP (2015) Ductilizing of a brittle as-cast hypereutectic Al-Si alloy by friction stir processing. Mater Lett 159:417–419. https://doi.org/10.1016/j.matlet.2015.07.006

14. Jana S, Mishra RS, Baumann JB, Grant G (2010) Effect of friction stir processing on fatigue behavior of an investment cast Al-7Si-0.6 Mg alloy. Acta Mater 58:989–1003. https://doi.org/10.1016/j.actamat.2009.10.015
15. Abubaker HM, Merah N, Al-Badour F et al (2021) Influence of friction stir processing on wear, corrosion, and fracture toughness behavior of 2507 super duplex stainless steel. J Mater Eng Perform 30:89–102. https://doi.org/10.1007/s11665-020-05325-4
16. Yang KV, Cáceres CH, Nagasekhar AV, Easton MA (2012) The skin effect and the yielding behavior of cold chamber high pressure die cast Mg-Al alloys. Mater Sci Eng A 542:49–55. https://doi.org/10.1016/j.msea.2012.02.029
17. Gourlay CM, Laukli HI, Dahle AK (2007) Defect band characteristics in Mg-Al and Al-Si high-pressure die castings. Metall Mater Trans A 38:1833–1844. https://doi.org/10.1007/s11661-007-9243-1
18. Otarawanna S, Laukli HI, Gourlay CM, Dahle AK (2010) Feeding mechanisms in high-pressure die castings. Metall Mater Trans A 41:1836–1846. https://doi.org/10.1007/s11661-010-0222-6
19. Jiao XY, Zhang YF, Wang J et al (2021) Characterization of externally solidified crystals in a high-pressure die-cast AlSi10MnMg alloy and their effect on porosities and mechanical properties. J Mater Process Technol 298. https://doi.org/10.1016/j.jmatprotec.2021.117299
20. ASTM standard, Designation: B871-01, Standard Test Method for Tear Testing of Aluminum Alloy Products (Current edition approved April 1, 2007. Published May 2007. Originally approved in 1996. Last previous edition approved in 2001 as B 871–01) (2007)

Effect of Friction Stir Processing on the Microstructure and Mechanical Properties of Thick Al-6061 Alloy

Amlan Kar, Eric J. Pickron, Todd Curtis, Bharat K. Jasthi, Wade Lein, Zackery McClelland, and Grant Crawford

Abstract This paper highlights the influence of severe thermomechanical processing conditions on fabricating a high-strength and tough, 12 mm thick aluminum alloy (Al-6061-T6). An optimum thermomechanical condition using friction stir processing (FSP) resulted in a material with better tensile properties and a fivefold increase in impact toughness. Detailed characterization revealed the evolution of a refined microstructure with a uniform distribution of second-phase particles. Grain size variation at different processing conditions is examined to identify the mechanisms of microstructural evolution. The tensile sample with 100% processed material fails at a location of refined microstructure and inhomogeneous deformation features. The improved tensile strength and toughness of the processed zone are attributed to homogeneous deformation, recrystallized microstructure evolution, grain refinement, and quality of processed zone to absorb energy, homogeneous distribution of precipitates, and location of the fracture in the stir zone.

Keywords Severe thermomechanical processing · Friction stir processing · Tensile properties · Impact toughness · Grain refinement · Continuous dynamic recrystallization

Introduction

In recent years, the demand for lightweight ballistic materials with better mechanical properties and performance has triggered the need for a new processing technology

A. Kar · E. J. Pickron · T. Curtis · B. K. Jasthi · G. Crawford (✉)
Arbegast Materials Processing and Joining Laboratory (AMP), South Dakota School of Mines & Technology, Rapid City, SD 57701, USA
e-mail: Grant.Crawford@sdsmt.edu

W. Lein
U.S. Army Corps of Engineers Cold Regions Research and Engineering Laboratory, Vicksburg, MS, USA

Z. McClelland
U.S. Army Engineer Research and Development Center, Vicksburg, MS, USA

Y. Hovanski et al. (eds.), *Friction Stir Welding and Processing XII*,
The Minerals, Metals & Materials Series,
https://doi.org/10.1007/978-3-031-22661-8_5

to modify base materials. Friction stir processing (FSP) is a promising solid-state technique that can be used to improve the mechanical properties (e.g., tensile and impact toughness) of a variety of metal alloys [1]. FSP was adapted from the concepts of friction stir welding where a non-consumable rotating tool plunges into the base material and traverses along the processing direction to induce severe thermomechanical deformation in the processing material. The severe plastic deformation process leads to grain refinement by dynamic recrystallization mechanism which result in improved mechanical properties of the processed material [2]. On the other hand, the evolution and distribution of precipitates play an important role in the final mechanical properties of the processed materials. Plastic deformation and the local thermal profile induced by the thermomechanical process affect the precipitation kinetics and precipitate fraction, which influence corresponding precipitation strengthening [2]. During FSP, the grain size and precipitate characteristics vary with feedback torque and heat input. Further, processing conditions vary depending on the extrinsic processing parameters, such as tool rotation speed, processing speed, tool tilt angle, and forging load. Consequently, variation in internal microstructure, local mechanical properties, and ultimately performance can arise if a steady-state condition is not achieved at optimized processing parameters. Nonuniform deformation owing to non-steady-state processing during FSP can make the material susceptible to the formation of defects, such as void, porosity, and groove defects [3]. The presence of defects promotes crack initiation, thus limiting strength, ductility, and toughness of the materials. It is important to engineer the deformation characteristics to minimize or eliminate defect formation during FSP. Deformation characteristics in FSP can be monitored by supervising feedback torque and application of steady-state forging load. Torque output has a direct correlation with heat input and the peak temperature of processing. Banik et al. [4] reported that higher torque and forging force can result in higher heat inputs. Higher heat input often results in thermal softening and improved material flow during processing, which can lead to establishment of a steady-state processing condition. However, higher heat input can also lead to higher peak temperature, longer cooling time, and abnormal grain growth irrespective of steady and non-steady conditions [5].

Hence, it becomes essential to employ a steady-state forging load to get defect-free material with superior grain refinement and mechanical properties. Superior mechanical properties can be achieved in bulk material with ultrafine grains and homogeneous distribution of precipitates produced by severe plastic deformation that can be achieved under steady-state and optimized processing conditions. Changing the steady-state forging load while maintaining other processing parameters (e.g., tool rotation speed) essentially varies the degree of thermomechanical condition for friction stir processing. There are a number of research articles correlating a particular processing parameter with corresponding mechanical properties without monitoring the steady-state condition of processing, which can often lead to increased property variation and the evolution of micro-defects in the processed zone [3]. The production of micro-defects in this instance is often due to non-steady-state deformation, and their presence often confounds mechanical property measurements because the inherent strength of the processed zone cannot be measured when micro-defects promote failure under external load. Hence, the true characteristics of processed materials under non-steady-state conditions cannot be obtained for comparison with that of other processing parameters.

Therefore, in the present investigation, the FSP of Al-6061 alloy at different tool rotation speeds was performed under steady-state processing conditions to vary severe thermomechanical deformation conditions. Microstructure evolution and mechanical properties (i.e., tensile strength, ductility, and impact toughness) were characterized to assess the suitability of FSP Al-6061 alloys for ballistic applications. Microstructure and mechanical properties are correlated with deformation conditions at different tool rotation speeds to determine the mechanism associated with the superior performance of the developed material at optimum steady-state processing conditions.

Experimental Procedure

Aluminum (Al) 6061-T6 plates (Al-6061) with dimensions of 240 × 120 × 12 mm were used as the starting material. Prior to FSP the plates were mechanically polished using scotch bite and cleaned with acetone. The mechanically cleaned plates were subjected to FSP in three different processing conditions shown in Table 1. Externally, tool rotation speeds and corresponding steady-state forging loads in load control mode were varied to compare the results. The FSP process tool was made of H13 tool steel with a frustum-shaped pin and spiral thread features. The shoulder diameter of the tool was 20 mm. The frustum-shaped pin had a height of 10 mm and top and bottom diameter of 6 and 4 mm, respectively. A schematic diagram of the processing is shown in Fig. 1. The image shows the locations of tensile, charpy, and metallographic (optical) samples in the processed plate.

After FSP, the plates were sectioned using the waterjet cutting. Specimens for microstructural examination were prepared using standard polishing techniques. Final polishing of the specimens was conducted using a 1.0 μm diamond paste followed by a 0.05 μm diameter colloidal silica suspension. The polished samples

Table 1 Processing conditions of three FSP experiments with different tool rotation speed and steady-state forging loads

Process Parameters	Condition 1	Condition 2	Condition 3
Tool rotational speed (rpm)	400	600	800
Processing speed (mm/min)	256	256	256
Tool tilt angle (°)	1	1	1
Steady-state forging load (kg)	2600	2200	1600
Torque feedback (Nm)	103	70	54

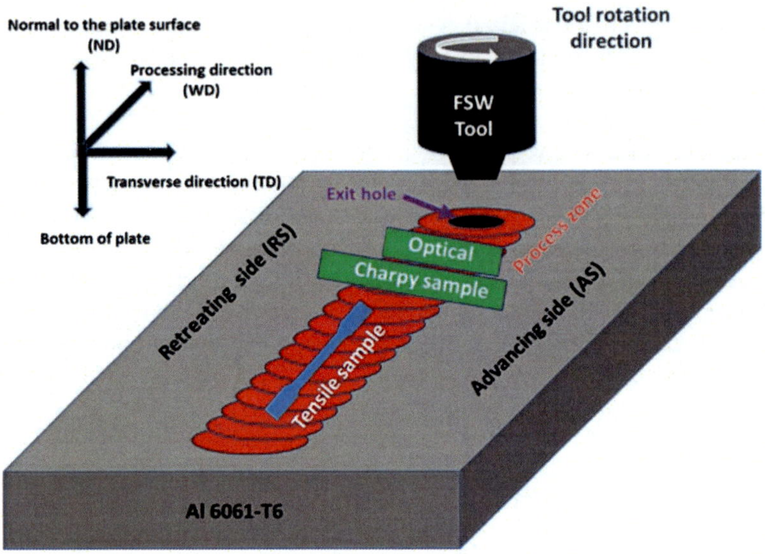

Fig. 1 Schematic diagram of the friction stir processing and extracted samples locations in the processed plate

were etched with Keller's reagent comprising 2.5 ml HNO_3, 1.5 ml HCl, 1 ml HF, and 95 ml distilled water. Microstructural examination was performed using optical and scanning electron microscopy (SEM). The average grain size was calculated using the line intercept method with five lines in a mutually perpendicular direction. Detailed microstructural and chemical analyses were achieved by using using a dual beam field emission SEM (Helios 5 CX, Thermo Scientific, Massachusetts, USA).

Tensile testing was performed to evaluate the mechanical properties of the base material (i.e., Al-6061) along the rolling direction and the longitudinal direction (processing direction, see Fig. 1) of the FSP material. Sub-size ASTM E8M tensile specimens with a thickness of 5.5 mm, an overall length of 90 mm, and a gauge length of 25 mm were sectioned using a waterjet cutting and milled to the required dimensions using a conventional milling machine. Tensile tests according to ASTM E8M standard were performed using a servohydraulic uniaxial testing system (Test Resources, Inc., Model 313, Shakopee, MN, USA) at room temperature with a crosshead speed of 1.27 mm/min.

Results and Discussion

Microstructure Characterization

The macrostructure of the processed materials fabricated at different tool rotation speed under different thermomechanical processing conditions is shown in Fig. 2. A transition of microstructure from base material to processed zone on the retreating side (RS) and advancing side (AS) can be observed. No significant difference in the process zone can be noticed. However, an inhomogeneous deformation zone (indicated by an arrow in Fig. 2a) is present in all samples. The features in this zone become coarser with increasing tool rotation to 600 rpm (Fig. 2b). At 800 rpm, the inhomogeneous deformation zone exhibits elongated light contrast regions which represent base metal flow features at the bottom-middle part of the pin-driven deformation zone (Fig. 2c). These local inhomogeneous deformation regions and deposition features (material flow) may have an effect on local stress concentration that results in the failure of the tensile sample.

The microstructural characteristics (e.g., grain size and second-phase particle size/distribution) varied depending on the location within the process zone. For comparison, optical micrographs obtained from the center of the process zone are provided in Fig. 3a–c while Fig. 3d shows the base metal (Al-6061) microstructure away from the FSP region. Clearly, significant grain refinement occurred in the processed zone for all processing conditions when compared to the base metal. Table 2 summarizes the grain size and aspect ratio for each specimen. Moreover, increased tool rotational speed and reduced steady-state forging load resulted in a reduction in grain size in the processed region. Consequently, the specimen produced at 800 rpm exhibited superior grain refinement (Fig. 3c and Table 2). The grains, however, were not homogeneous in morphology as the aspect ratio deviates from 1.0 at all processing conditions. The aspect ratio is defined as grain size measured along the shortest axis divided by the longest axis of the microstructure. The microstructure of the specimen produced at 600 rpm exhibited the lowest aspect ratio (i.e., 0.67) among all the processed materials. This could be attributed to the combination of medium torque output and medium tool rotation speed, which delay the dynamic recrystallization for grain evolution. At a lower tool rotation speed of 400 rpm and higher forging

Fig. 2 Macrostructure of the processed materials fabricated at different tool rotation speeds of **a** 400 rpm, **b** 600 rpm, and **c** 800 rpm showing a difference in materials flow lines at steady-state processing conditions

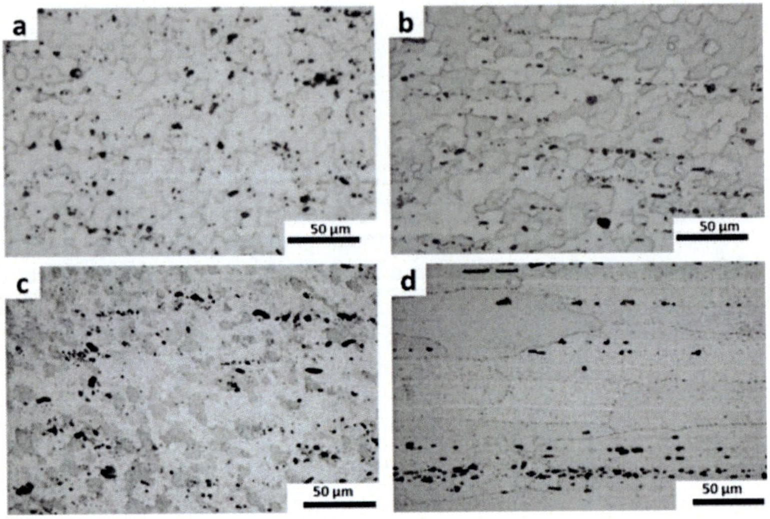

Fig. 3 Optical microstructure showing grain morphology of the process zone at different tool rotation speeds of **a** 400 rpm, **b** 600 rpm, and **c** 800 rpm at steady-state processing conditions. **d** Base Al-6061 microstructure exhibits coarse and elongated grains

Table 2 Measured grain size and aspect ratio for the processed materials

Tool rotation speed (rpm)	Grain size (μm)	Aspect ratio
400	16.2 ± 2.6	0.80 ± 0.11
600	14.9 ± 3.3	0.70 ± 0.02
800	13.0 ± 1.6	0.82 ± 0.09
Base Al-6061	79.3 ± 19.8	0.57 ± 0.06

load, Al is subjected to severe deformation (highest torque in Table 1) that promotes strain-induced continuous dynamic recrystallization [1, 6]. On the other hand, FSP at 800 rpm leads to deformation of materials at lower forging load leading to less severity in deformation and homogeneous post-process grain growth. Hence, more homogeneous grain evolution with the highest aspect ratio (ratio = 0.82 in Table 2) has taken place at 800 rpm, as measured in Table 2.

Microchemical Analysis of the Process Zone

Figure 4 shows an SEM image and corresponding EDS elemental maps obtained from the center of the processed zone for the 800 rpm specimen. A number of dark precipitates (Fig. 4a) exhibiting a relatively uniform distribution can be observed in this region. From inspection of the EDS elemental maps (Fig. 4b–f), Mg, Si, and Fe are enriched in regions correlating with the dark precipitates indicating the likely

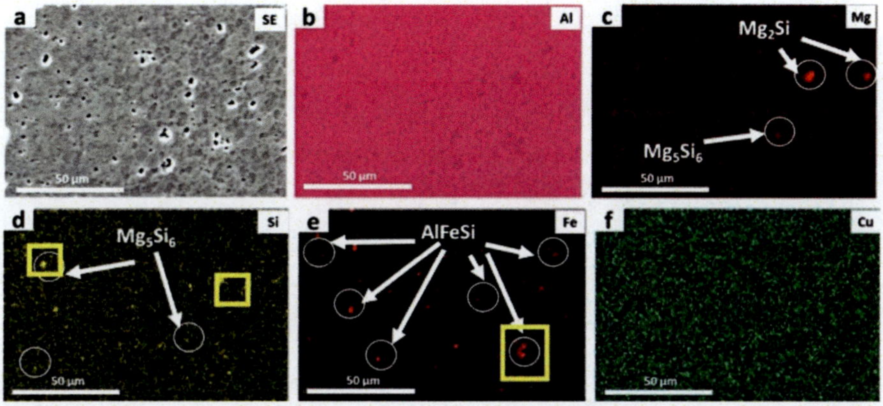

Fig. 4 SEM–EDS maps showing the precipitates and their distribution in the processed zone with 800 rpm

presence of conventional Al-6061 precipitates including Mg_2Si (T-Phase, encircled in Fig. 4c) [7], Mg_5Si_6, (encircled in Fig. 4c–d) [8], and AlFeSi (encircled in Fig. 4e). It is important to note that these are likely precipitates based on current EDS analysis. Additional analysis would be needed to confirm their identity.

Compared with the base Al-6061, they seem overaged due to higher homologous processing temperature and exposure time (time for processing) (marked with square boxes in Fig. 4d–e). Hence, a few coarse precipitates of size more than 5 μm can be seen in the Mg map (Fig. 4c). Similar coarse precipitates were reported by Kalinenko et al. [9]; the authors further stated that the reduction in mechanical properties of the stir zone was attributed to the reduction in precipitates due to the phase transformation and reduction in dislocation density. It is important to report that the processing temperature could reach up to 500 °C [5]. The uniform distribution of the precipitates promotes homogeneous deformation in mechanical testing leading to improvement in mechanical properties. All materials developed here are subjected to deformation by the same route and at identical peak temperatures leading to recrystallization. Hence, identical composition is expected to evolve at different tool rotation speeds. There may be a little difference in the size of the precipitates at different thermomechanical conditions due to differences in heat input and rate of deformation at a different tool rotation speed.

Tensile Properties

Tensile properties of processed materials at three different processing conditions are shown in Fig. 5. The properties provided in Fig. 5 are reflective of the mechanical properties of the FSP region only as the tensile samples were extracted along the processing direction and within the stir zone.

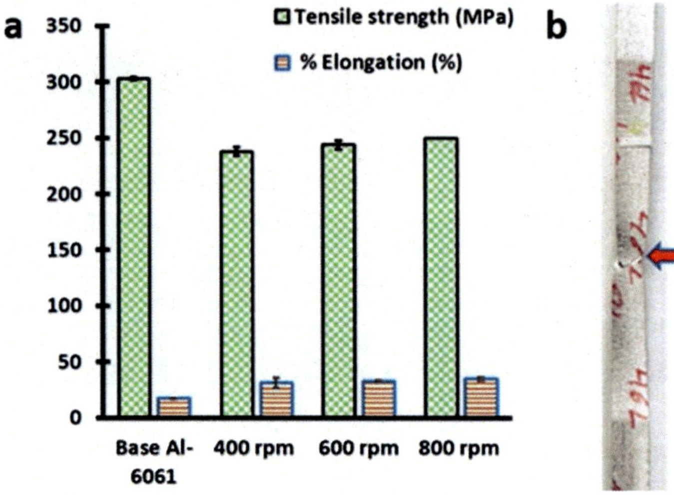

Fig. 5 **a** Longitudinal tensile properties of the welds with different processing conditions and **b** location of fracture for the tensile sample with 800 rpm and steady-state forging load of 1600 kg

Table 3 provides the measured ultimate tensile strength (TS) and percent of elongation (ductility) for each specimen as well as the joint efficiency (ratio of processed material TS to base metal TS). Considering the measured variability, a significant variation in either TS or ductility was not observed with varied FSP conditions. The smaller variation in properties indicates that tool rotation speed has little influence on tensile properties. From the micrographs of the fractured tensile specimens shown in Fig. 5b, it can be decided that the tensile specimen failed in center of the processed zone. This pointed to the homogeneous deformation of metal due to the evolution of recrystallized microstructure with weak crystallographic texture in the stir zone [10, 11]. Even though the process zone contains a variation in microstructure due to recrystallization at top and inhomogeneous deformation at the bottom of the process zone, slow rate of deformation minimizes the strain concentration leading to more homogeneous deformation. The higher tensile properties could be attributed to recrystallized microstructure, grain refinement, homogeneous distribution of precipitates, location of fracture in the stir zone.

Table 3 Quantitative longitudinal tensile properties with standard deviation for FSP of Al-6061-T6 at three different processing conditions

Sample ID	Tensile strength (MPa)	% Elongation	Joint efficiency (%)
Base Al-6061	303 ± 1.2	17.3 ± 0.31	100
400 rpm Longitudinal	238 ± 3.7	31.4 ± 4.86	79
600 rpm Longitudinal	244 ± 3.9	33.3 ± 0.55	81
800 rpm Longitudinal	250 ± 1.8	35.3 ± 1.15	83

SEM images of the fracture surfaces of the processed and base Al samples are shown in Fig. 6. It is important to note that fracture location in all the processed samples consisted of refined microstructure, a high fraction of grain boundaries [12] due to continuous dynamic recrystallization of Al-6061 and evidence of weak crystallographic texture [5]. From Fig. 6a–c, all the FSP samples exhibited tear ridges (indicated by white box). However, the fractional quantity of tear ridges varies with processing conditions. For example, a low fraction of tear ridges is shown in the weld with 400 rpm (Fig. 6a) compared to that with 600 rpm (Fig. 6b) and 800 rpm (Fig. 6c). On the other hand, the dimples' size is finer in the weld with 800 rpm (indicated by white arrow). Yousefpour et al. [13] reported that the evolution of tear ridges corresponds to crystallographic texture, whereas the grain size of the dimples depends on the initial grain size of the tensile specimen. The images predict that the weld with 800 rpm has refined microstructure and higher crystallographic texture compared to other FSPed specimens and base Al (Fig. 6d). In addition, the presence of tear ridge can be correlated with the inhomogeneous deformation features seen in Fig. 2. The presence of tear ridge increases with an increase in deformation features and tool rotation speed.

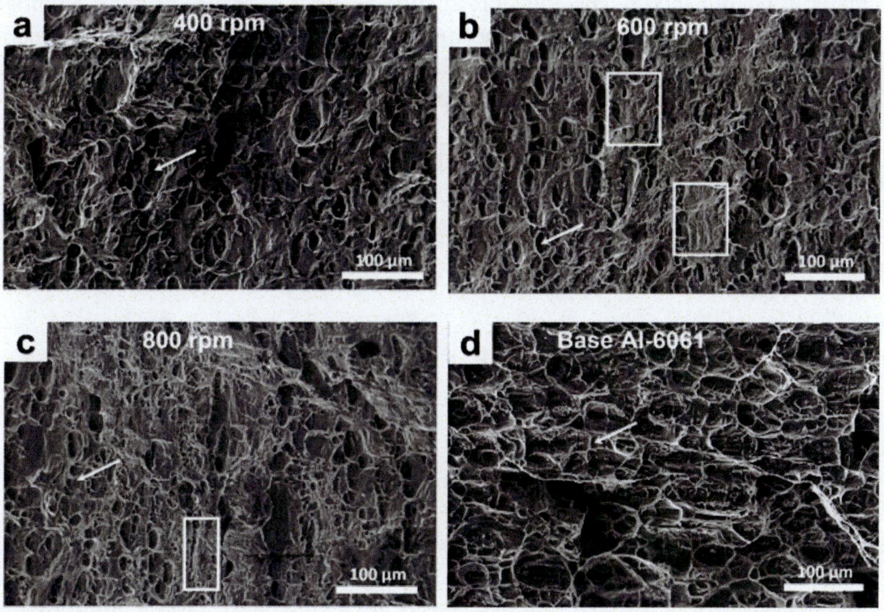

Fig. 6 SEM fractographs of the **a–c** processed tensile samples and **d** base Al-6061 showing different in dimple sizes and morphologies with variation in processing conditions. Dimple and tear ridge are indicated by white arrow and white box

Table 4 Charpy impact toughness for FSP samples and base Al-6061

Sample information	Impact toughness (Joule/m)
FSP with 400 rpm	76 ± 2.8
FSP with 600 rpm	57 ± 1.9
FSP with 800 rpm	56 ± 1.1
Base Al-6061	15 ± 1.3

Improvement in Impact Toughness

Mechanical tests, performed at a speed of 1.27 mm/min, cannot replicate conditions of high impact deformation behaviour of a material. Impact toughness for V-notch samples were used to determine the resistance of material to penetration under high-velocity impact. The impact toughness values for all processed materials are shown in Table 4. The FSP sample with 400 rpm has greater impact energy than all other materials. It exhibits a five-time improvement in impact toughness compared to that of base Al-6061. In impact loading, the energy of the applied load is ultimately absorbed and transferred to the adjacent material to resist the load. The processed material having inhomogeneous deformation features observed at the bottom side of the process zone (Fig. 2) acts as a stress concentrator due to insufficient time for load transfer. The surrounding recrystallized microstructure absorbs higher energy and accommodates a higher load. The stress discontinuity across this zone leads to the formation and propagation of cracks leading to failure of the V-notch Charpy sample. The increased severity of the inhomogeneous deformation region observed at 800 rpm (Fig. 2c) is prone to easy fracture during impact testing and hence shows lower impact toughness. On the other hand, the superior impact toughness at 400 rpm reveals that the processed materials have significant energy absorbing capacity under impact load.

Conclusion

Friction stir processing of aluminum 6061-T6 alloy at different thermomechanical conditions with the variation of tool rotation speed and corresponding variation in steady-state forging loads has been carried out. Microstructure characterization, chemical analysis, and mechanical properties evolution (tensile and impact toughness) were performed in detail. From the results and analysis, the following conclusions can be drawn.

1. Steady-state severe thermomechanical deformation at different tool rotation speeds results in defect-free welds, although there is a substantial variation in feedback torque value. A variation of material flow at the bottom side (tool pin-affected zone) of the processed zone appears due to relative variation in pin-driven materials flow and inhomogeneous deformation at different tool rotation speeds.

At lower rotation and higher rotation, the flow line seems to be homogeneous special dots and discontinuous elongated deposition lines, respectively, due to possible variation in recrystallization and evolution of crystallographic texture.
2. Chemical analysis from the center-processed zone reveals the evolution of Mg, Si, and Fe base precipitates. The evolution of coarse precipitates (overaging) and microstructure with an aspect ratio of 1.0 illustrate higher homologous processing temperature and extended exposure time; this leads to a reduction in tensile strength and improvement in ductility and impact toughness at all processing conditions.
3. The tensile sample with 100% processed material fails at the center of the processed zone that contains refined microstructure. There is a little improvement (less than 5% overall variation) in tensile properties with an increase in tool rotation speed. This is attributed to recrystallized microstructure developed in the processed zone, which minimizes the deformation strain leading to homogeneous deformation at low speed of deformation (crosshead speed of 1.27 mm/min) even though there is substantial variation in microstructure evolution from top to bottom of the processed zone. The higher tensile properties could be attributed to recrystallized microstructure, grain refinement, homogeneous distribution of precipitates, and location of the fracture in the stir zone.
4. A substantial difference in impact toughness in the range of 25% can be noticed with the change in tool rotation speed and feedback torque due to variation in inhomogeneous deformation features at the bottom side (pin-driven flow zone) of the processed zone. A five-time improvement in impact toughness at 400 rpm in comparison with base Al-6061 alloy has been reported due to the reduction in inhomogeneous feature in the processed zone and evolution of recrystallized microstructure, which minimizes stress contraction and absorbs higher impact energy.

Acknowledgements Funding for this research was supported by the US Army Engineer Research and Development Center. The use of trade, product, or firm names in this document is for descriptive purposes only and does not imply endorsement by the U.S. Government. The tests described and the resulting data presented herein, unless otherwise noted, are based upon work supported by the US Army ERDC under PE 0603119A, "Ground Advanced Technology ", Task" Materials and Manufacturing Technology for Cold Environments. Permission was granted by the Director, Cold Regions Research and Engineering Laboratory to publish this information. The findings of this paper are not to be construed as an official Department of the Army position unless so designated by other authorized documents.

References

1. Kumar S, Kar A (2021) A review of solid-state additive manufacturing processes. Trans Indian Nat Acad Eng 6(4):955–973

2. Samanta A, Seffens RJ, Das H et al (2022) Microstructure-refinement-driven enhanced tensile properties of high-pressure die-cast A380 alloy through friction stir processing. J Manuf Process 78:352–362
3. Al-Moussawi M, Smith AJ (2018) Defects in friction stir welding of steel. Metallogr Microstruct Anal 7(2):194–202
4. Banik A, Saha Roy B, Deb Barma J et al (2018) An experimental investigation of torque and force generation for varying tool tilt angles and their effects on microstructure and mechanical properties: Friction stir welding of AA 6061–T6. J Manuf Process 31:395–404
5. Kar A, Kailas S, Suwas S (2020) Mechanism of variation in high-temperature grain stability of aluminum in dissimilar friction stir welds. https://doi.org/10.1520/MPC20190011
6. Kar A, Yadav D, Suwas S et al (2020) Role of plastic deformation mechanisms during the microstructural evolution and intermetallics formation in dissimilar friction stir weld. Mater Charact 164:110371
7. Zeng C, Ghadimi H, Ding H et al (2022) Microstructure evolution of Al6061 alloy made by additive friction stir deposition. Mater (Basel, Switzerland) 15(10):3676
8. He T, Shi W, Xiang S et al (2021) Influence of aging on corrosion behaviour of the 6061 cast aluminium alloy. Mater (Basel, Switzerland) 14(8):1821
9. Kalinenko A, Kim K, Vysotskiy I et al (2020) Microstructure-strength relationship in friction-stir welded 6061-T6 aluminum alloy. Mater Sci Eng A 793:139858
10. Hart EW (1967) Theory of the tensile test. Acta Metall 15(2):351–355
11. Kar A, Kailas SV, Suwas S (2020) Mechanism of variation in high-temperature grain stability of aluminum in dissimilar friction stir welds. Mater Perform Character 9(2):262–271
12. Kar A, Suwas S, Kailas SV (2019) Multi-length scale characterization of microstructure evolution and its consequence on mechanical properties in dissimilar friction stir welding of titanium to aluminum. Metall Mater Trans A 50:5153–5173
13. Yousefpour F, Jamaati R, Aval HJ (2021) Effect of traverse and rotational speeds on microstructure, texture, and mechanical properties of friction stir processed AZ91 alloy. Mater Charact 178:111235

Graphite-Reinforced 6201 Aluminum Alloy Fabricated by In-Situ Friction Stir Processing: Process, Microstructure, and Mechanical/Electrical Properties

Yijun Liu, Gaoqiang Chen, Fangzheng Shi, Mengran Zhou, Shuai Zhang, Gong Zhang, and Qingyu Shi

Abstract Friction stir processing (FSP) has proved to be successful in the past two decades for the preparation of novel metal matrix composites with preferable properties. In this work, graphite was added into 6201 aluminum alloy through multi-pass in-situ FSP in order to fabricate graphite-reinforced 6201 aluminum alloy. We look into the microstructure, the mechanical, and electrical properties of the composites. It is found that, after multi-pass FSP, the graphite is significantly fragmented and uniformly distributed in the aluminum matrix. The tensile strength and electrical conductivity of the composites with 9-pass FSP are improved by 19.5 and 1.2% in comparison with the matrix. The number of FSP passes is an important parameter affecting the microstructure of the composites. Effect of the number of FSP passes on the tensile strength and electrical conductivity of the composites is investigated.

Keywords Graphite · Composite · Tensile strength · Electrical conductivity

Introduction

Metal materials play an extremely important role in modern social life as a transmission medium for electric power. Mechanical and electrical properties are two of the most important properties for metal materials [1–4]. The improvement of mechanical and electrical properties of metal materials is of great significance to the cable industry and the field of power transmission.

However, for a long time, there has been a contradiction between the mechanical properties and electrical properties of metal materials [1, 2, 5]; that is, the improvement of one performance is usually accompanied by the decrease of the other performance [5, 6]. The contradiction between the mechanical properties and electrical properties of metal materials has become a major limitation to resource conservation and technological progress. Therefore, it is of great importance to overcome

Y. Liu · G. Chen · F. Shi · M. Zhou · S. Zhang · G. Zhang · Q. Shi (✉)
Department of Mechanical Engineering, Tsinghua University, Beijing 100084, China
e-mail: shqy@tsinghua.edu.cn

Y. Hovanski et al. (eds.), *Friction Stir Welding and Processing XII*,
The Minerals, Metals & Materials Series,
https://doi.org/10.1007/978-3-031-22661-8_6

the contradiction between the two properties to improve the strengthening mechanism of mechanical and electrical properties of metal materials and to promote their application in modern society.

In this paper, graphite-reinforced high-conductivity aluminum alloy composites were prepared by friction stir processing. The mechanical and electrical properties of the composites were tested, the microstructure of the composites was characterized, and the mechanism of the influence of processing passes on the mechanical and electrical properties of the composites was discussed.

Experimental Procedure

The matrix material used for the preparation of the composites was rolled 6201 aluminum alloy plates with length of 300 mm, width of 150 mm and thickness of 6 mm. The main element composition of the aluminum alloy is shown in Table 1. The reinforcing phase of the composites is graphite nano-powder with a particle size of 3–6 μm and a thickness of less than 500 nm. The secondary electron image of the graphite powder under SEM is shown in Fig. 1a. The graphite is flaky and has a relatively complete crystal structure. The schematic diagram of the process of fabricating composites by FSP is shown in Fig. 1b. A groove of 240 mm in length, 1 mm in width, and 4 mm in depth was machined on the aluminum plate by CNC milling machine, and the graphite powder was filled into the groove. Firstly, a 20 mm diameter stirring head without a needle was used to cover the groove to prevent the graphite powder from escaping during the subsequent processing, with a rotating speed of 1500 rpm and a welding speed of 60 mm/min. Then, a cylindrical right threaded stirring head with a needle length of 5 mm, a needle diameter of 5 mm and a shoulder diameter of 20 mm was used to process 3-passes and 9-passes of FSP along the groove, with a rotating speed of 1000 rpm and a welding speed of 60 mm/min. Composites with graphite content of 0.1 wt.% were prepared, as shown in Fig. 1c. The 3-pass-FSPed composites, 9-pass-FSPed composites, and the base metal (BM) were subjected to T6 heat treatment (510 °C, 1 h solution treatment followed by water quenching, and then 175 °C, 8 h artificial aging treatment). All the samples of composites were selected in the stir zone.

The SEM samples were ground on 400#, 800#, and 1000# SiC papers, respectively, and then mechanically polished. Al_2O_3 was chosen as the polishing agent to avoid external C elements. Since corrosion may lead to the exfoliation of the reinforcement, the samples polished to a mirror effect were directly observed under SEM. A Zwick Z2.5 TH universal testing machine was used for tensile testing of the samples, and

Table 1 Chemical composition of the AA6201 alloy (wt.%)

Mg	Si	Fe	Ti	Mn	Zn	Cr	Cu	Al
0.62	0.53	0.17	0.03	0.02	0.002	0.002	0.0001	Bal

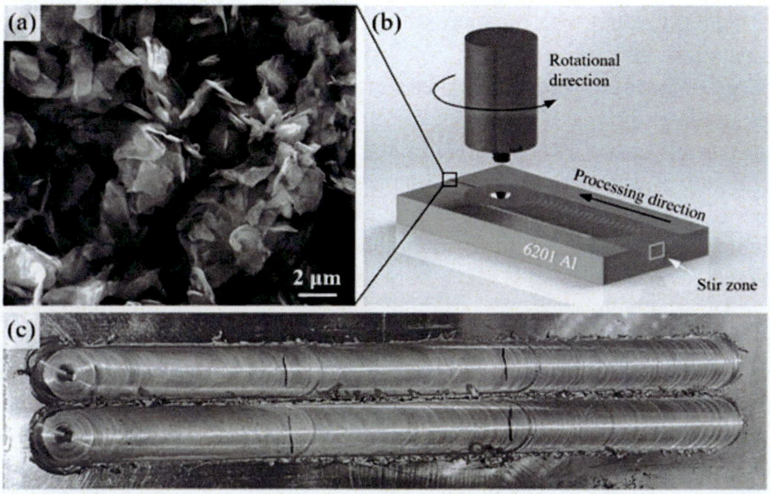

Fig. 1 **a** SEM image of graphite nano-powder; **b** Schematic of FSP; **c** Morphology of the graphite reinforced composites

a four-point method was used for electrical conductivity testing of the samples. The length direction of the samples was parallel to the FSP direction to obtain samples with uniform composition, and three replicate samples were prepared under each same parameter to test their mechanical and electrical properties.

Results and Discussion

Figure 2 shows the secondary electron images of the composites under SEM, in which Fig. 2a, b shows the 3-pass-FSPed composite and Fig. 2c, d shows the 9-pass-FSPed composites. In different samples, the graphite reinforcement is tightly embedded in the aluminum matrix without obvious interfacial defects, and the graphite is uniformly distributed without agglomeration. Different from the complete flake morphology of pristine graphite, the morphology of the graphite reinforcement in the composites is a stacked fish scale shape, as shown in Fig. 2b, d, which may be caused by the severe plastic deformation during FSP. In Fig. 1a, it shows that the particle size distribution of the pristine graphite is 3–6 μm. And, Fig. 2a shows that the particle size of the graphite reinforcement in the 3-pass-FSPed composite decreased to 0.1–1 μm. After 9-pass FSP, as shown in Fig. 2c, the size distribution of the graphite reinforcement is also 0.1–1 μm, but the reinforcement is smaller in average size, higher in density, and distributed with a smaller average distance, which indicates that in the process of FSP preparation of composites the severe plastic deformation mixes the graphite reinforcement with the aluminum matrix and breaks the graphite at the same time, and with the increase of processing passes, the degree of graphite fragmentation will also increase.

The graphite-reinforced composites were observed under TEM, as shown in Fig. 3. Figure 3a shows the dark-field (DF) image, the graphite reinforcement is light in color and the particle size is less than 500 nm, which is consistent with the size observed by SEM. Figure 3b shows the HADDF image; the image contrast represents the atomic number of the element. Since the atomic number of C element is lower than that of Al element, the graphite reinforcement shows dark color in the matrix. Figure 3c shows the EDS mapping of C element in the same area red indicates the C element-enriched area, which is consistent with the position of the reinforcement in Fig. 3a, b, proving that the reinforcement is indeed graphite. Figure 3d is a bright field image of a graphite reinforcement in the composite, and the HRTEM image at the position of the dotted box is shown in Fig. 3e. In Fig. 3d, the graphite reinforcement has a tight and defect-free combination with the aluminum matrix, and the interface is clean without the formation of compounds such as Al_4C_3. From Fig. 3e, the carbon atomic layer structure inside the graphite reinforcement can be seen. The interlayer spacing is 0.345 nm, which is close to the theoretical interlayer spacing of (0001) crystal plane of 0.335 nm. The morphology of the graphite reinforcement in Fig. 3d shows that the graphite reinforcement is not a complete single crystal structure, but consists of many multilayers of carbon atoms stacked together, which is also consistent with the scale-like structure observed in Fig. 2b, d. The effect of FSP on graphite is not

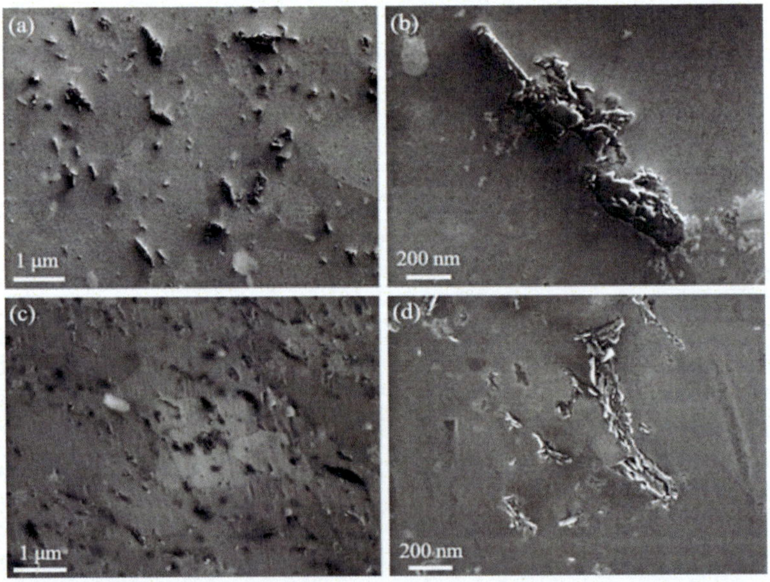

Fig. 2 SEM image of **a** the 3-pass-FSPed composites and **b** its graphite reinforcement; **c** the 9-pass-FSPed composites and **d** its graphite reinforcement

only the reduction of size, but also the change of structure, which is likely to have a corresponding impact on the physical properties of the material.

Tensile experiments were performed on the BM and the composites, and the fracture morphology was observed under SEM, as shown in Fig. 4. Figure 4a, b shows the tensile fracture morphology of the BM. The dimples in the fracture are elliptical, and the diameter of the dimples is about 5–20 μm. There is no second phase in the dimples, and the fracture morphology shows typical plastic fracture characteristics. Figure 4c, d shows the tensile fracture morphology of the 3-pass-FSPed composite. The size of the dimples is significantly smaller than that of the BM, about 1–2 μm, and flaky second phases are often seen at the bottom of dimples. The size of the flaky second phases tend to be larger in the larger dimples and smaller in the smaller dimples. The shallow depth of dimples indicates that the plasticity of the composite is lower than the BM. Figure 4e, f shows the tensile fracture morphology of the 9-pass-FSPed composite, the dimple size is further reduced to about 0.5–1 μm, and the dimple size is more uniform than the 3-pass-FSPed composite, and the size of the flaky second phases in the dimple is smaller.

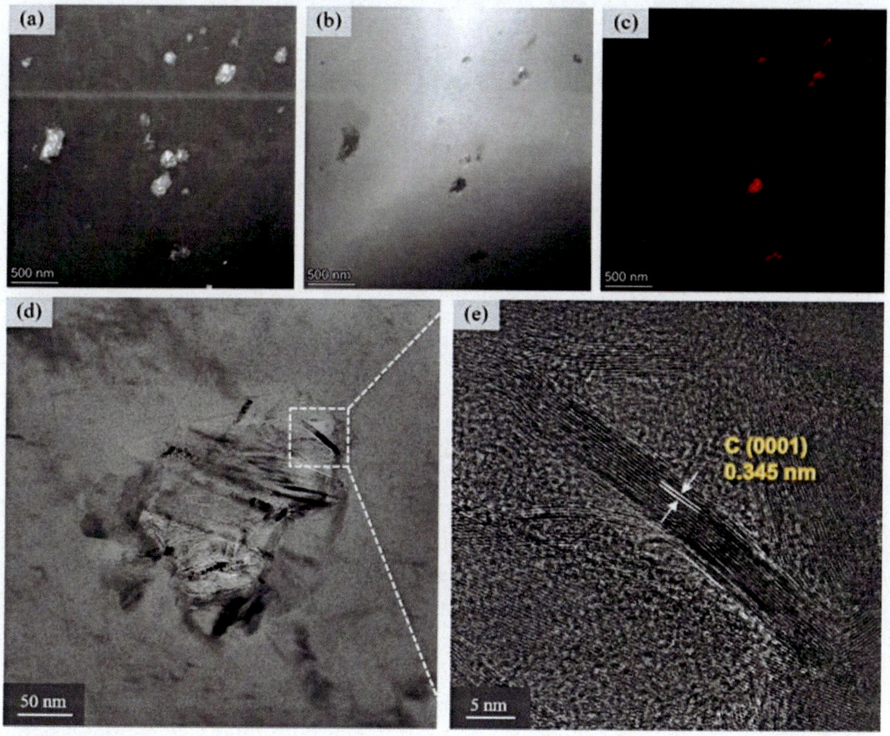

Fig. 3 TEM images of graphite-reinforced composites in **a** DF mode and **b** HADDF mode; **c** EDS mapping of C element; **d** bright field morphology of the graphite reinforcement in the composite, and **e** locally magnified HRTEM image

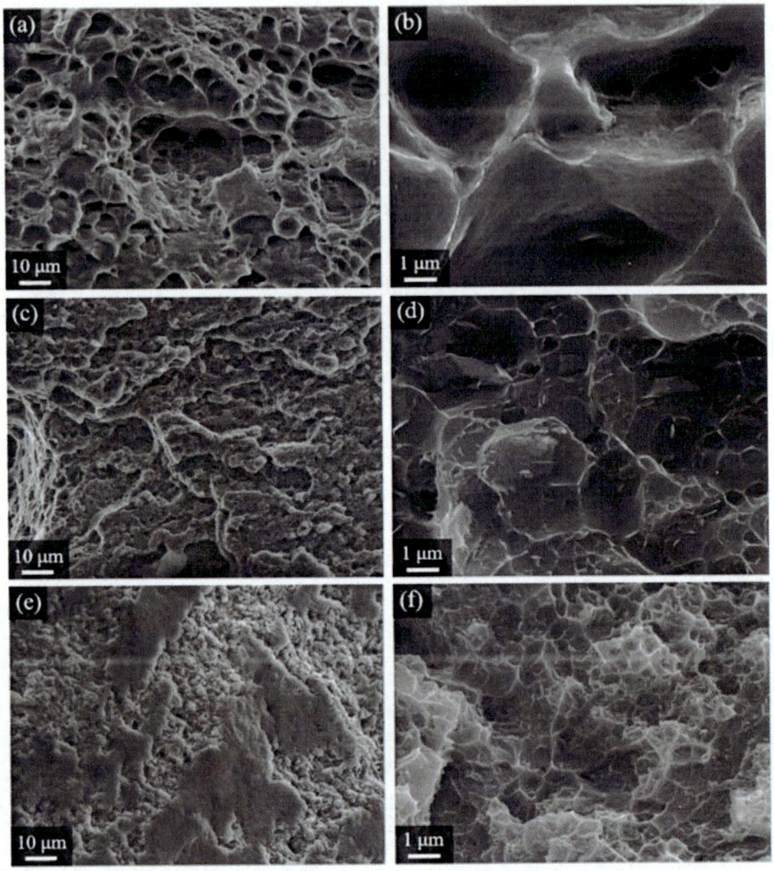

Fig. 4 SEM images of the fracture morphology of: **a, b** the BM; **c, d** the 3-pass-FSPed composite; **e, f** the 9-pass-FSPed composite

The elemental composition of the flaky second phases in the dimples is analyzed, as shown in Fig. 5. Dot 1 is selected on the flaky second phase, and dot 2 is selected on the matrix. The results are shown in Table 2. The elemental content of C at dot 1 is 71%, which is much higher than that at dot 2, indicating that the flaky second phases are graphite. The size of the graphite reinforcement in the dimples of the 9-pass-FSPed composite is smaller than that of the 3-pass-FSPed composite in the results in Fig. 4, which is also consistent with the results in Fig. 2. During the tensile fracture of the composites, reinforcement is often the nucleation site of dimples. Due to the increase of processing passes, the size of the graphite reinforcing phase decreases and the number increases, so the number of dimples also increases and the size decreases.

The tensile strength, elongation and electrical conductivity of the BM, 3-pass-FSPed composite and 9-pass-FSPed composite were tested, respectively, and the

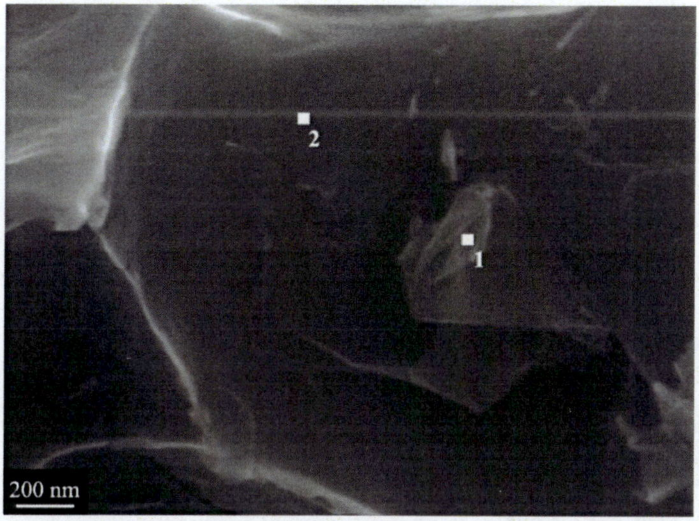

Fig. 5 SEM images of the graphite reinforced composite, white dots indicate locations where elemental composition testing was performed

Table 2 Element content at the position of the white dots in Fig. 5

Dot 1					Dot 2			
Element	Al	C	O	Mg	Element	Al	C	Si
At%	27.17	70.59	1.99	0.25	At%	55.47	44.12	0.42

results are shown in Fig. 6. The tensile strength of the BM is 263 MPa, the electrical conductivity is 47.8% IACS, and the elongation is 33%. Compared with the BM, the properties of the 3-pass-FSPed composite decreased slightly in all aspects, in which the tensile strength decreased to 215 MPa, the electrical conductivity decreased to 47.2% IACS, and the elongation decreased to 19%. However, as the number of processing passes increased to 9, the properties of the composite showed completely different results. The tensile strength and electrical conductivity of the 9-pass-FSPed composite were both improved compared with the BM. The tensile strength of the 9-pass-FSPed composite is 314 MPa, which is 19.5% higher than that of the BM, and the electrical conductivity is 48.3%IACS, which is 1.2% higher than that of the BM. Although the elongation is reduced to 25%, which is lower than that of the BM, it is also 30.4% higher than that of the 3-pass-FSPed composite. In the previous SEM results, it has been shown that with the increase of processing passes, the size of the graphite reinforcement phase becomes finer and the number becomes more, which may be the main factor contributing to the enhancement of the material properties with the increase in processing passes. Moreover, the error bars of the performance results of the 3-pass-FSPed composite, especially the error bar of the elongation results, is significantly longer than those of the BM and the 9-pass-FSPed composite,

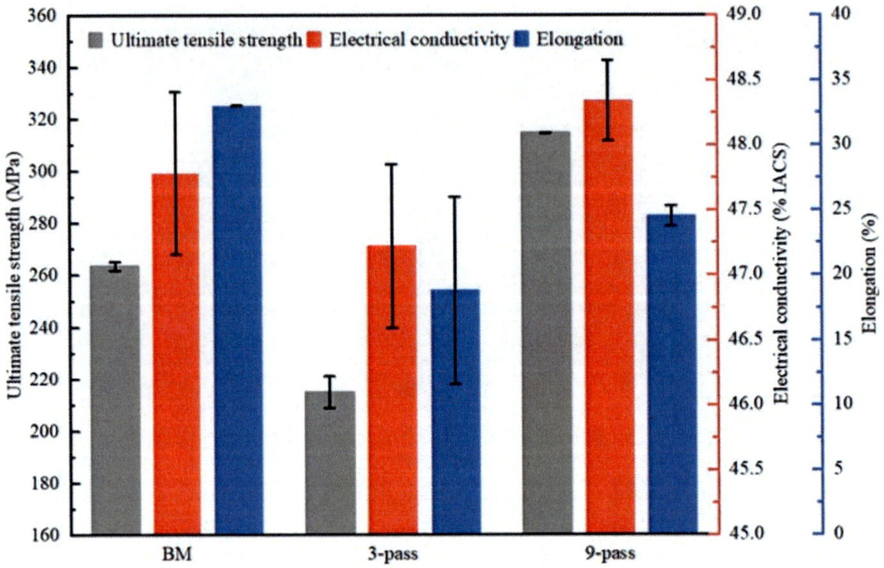

Fig. 6 Tensile properties and electrical conductivity of the composites and the BM

indicating that the performance results of the 3-pass-FSPed composite fluctuate more between different samples. This may be caused by the less uniform distribution of the graphite reinforcement on the macroscopic scale due to fewer processing passes, thus making the distribution density of the graphite reinforcement in different samples not identical.

One notable point in the performance results is that the mechanical–electrical properties of the graphite-reinforced composite samples after 9 passes of FSP are simultaneously improved compared to the base metal. In general metal processing technology, it is often difficult to improve the mechanical properties and electrical properties of materials at the same time [7–11]. In most cases, the processing process of strengthening the matrix by adding reinforcing phase can improve the tensile strength and at the same time significantly reduce the conductivity and elongation of materials. In recent studies, some scholars have used carbon nanomaterials such as graphene and carbon nanotubes as reinforcement to fabricate metal matrix composites with simultaneously improved mechanical–electrical properties [12–15], which is attributed to the excellent intrinsic mechanical–electrical properties of carbon nanomaterials. Graphite has the same intralayer atomic structure as graphene, so it also has excellent electrical and mechanical properties along the direction of the atomic layer. With the increase of processing passes, the micron-scale graphite is broken into submicron-scale and nano-scale, and its physical properties in all aspects will inevitably approach those of graphene from the physical properties of graphite. Therefore, the increase of the number of passes not only makes the distribution of the graphite reinforcement more uniform, finer and more numerous, but also further enhanced its intrinsic mechanical–electrical properties. The reduction in the size of the reinforcement, the increase in the number, and the improvement of its intrinsic physical properties work together, resulting in an increase in the comprehensive properties of the composites with the increase of processing passes.

Conclusion

In this study, graphite-reinforced aluminum matrix composites were fabricated by in-situ friction stir processing with three-pass and nine-pass. The microstructure and mechanical–electrical properties were investigated, respectively. SEM results show that in the composites of different passes the graphite reinforcement is closely embedded and uniformly distributed in the aluminum matrix, and with the increase of the pass, the size of the graphite reinforcement decreases and the density increases; TEM results show no defects in the interface between graphite reinforcement and matrix and no interfacial compounds. The tensile strength, electrical conductivity, and elongation of the 3-pass-FSPed composite are lower than those of the base metal, but the tensile strength and electrical conductivity of the 9-pass-FSPed composite are higher than those of the base metal, with an increase of 19.5% and 1.2%, respectively, and still had a high elongation of 25%. The improvement of composite properties with processing passes is considered to be the result of the combined effect of the reduction in the size and the increase in the density of the reinforcement.

References

1. Valiev RZ, Murashkin MY, Sabirov I (2014) A nanostructural design to produce high-strength Al alloys with enhanced electrical conductivity. Scr Mater 76:13–16
2. Cui X, Wu Y, Zhang G et al (2017) Study on the improvement of electrical conductivity and mechanical properties of low alloying electrical aluminum alloys. Compos B 110:381–387
3. Mondolfo LF (1976) Aluminum alloy: structure and properties
4. Luan BG, Tang XL, Huang CQ (1984) Increasing the electrical conductivity of aluminum conductor by treating the melt with boron
5. Lu L, Shen et al (2004) Ultrahigh strength and high electrical conductivity in copper. Science 304:422–426
6. Cui X, Wu Y, Liu X et al (2015) Effects of grain refinement and boron treatment on electrical conductivity and mechanical properties of AA1070 aluminum. Mater Des 86:397–403
7. Han SZ, Choi EA, Lim SH et al (2020) Alloy design strategies to increase strength and its trade-offs together. Prog Mater Sci 117:100720
8. Bass J (1972) Deviations from Matthiessen's rule. Adv Phys 21(91):431–604
9. Desai PD, James HM, Ho CY (1984) Electrical resistivity of aluminum and manganese. J Phys Chem Ref Data 13(4):1131–1172
10. Schafler E, Sachslehner F (2000) The electrical dislocation resistivity and the deviation from Matthiessen's rule of deformed high-purity aluminum. J Phys Condens Matter 121(50):10499–10514
11. Alley P, Serin B (1959) Deviations from Matthiessen's rule in aluminum, tin, and copper alloys. Phys Rev 116(2):334–338

12. Wang GJ, Ma YJ, Cai YP et al (2019) Overcoming the strength-conductivity trade-off dilemma in carbon nanotube/aluminum-copper fibers by diffusion interface and chemical reaction interface. Carbon 146:293–300
13. Ming Y, Lin W, Han Z et al (2017) Simultaneously enhancing the strength, ductility and conductivity of copper matrix composites with graphene nanoribbons. Carbon 118:250–260
14. Zhang S, Chen G, Qu T et al (2020) Simultaneously enhancing mechanical properties and electrical conductivity of aluminum by using graphene as the reinforcement. Mater Lett 265:127440
15. Zhang S, Chen G, Qu T et al (2021) A novel aluminum-carbon nanotubes nanocomposite with doubled strength and preserved electrical conductivity. Nano Res 14(8):2776–2782

Part III
Modeling and Validation

Analysis of Torque Data from Friction Stir Welds in Aluminum Alloys

Kevin Colligan

Abstract In the present work, welding procedures for friction stir welding (FSW) of aluminum alloys were collected from published and unpublished reports, permitting analysis of a diverse collection of alloys, material thickness, tool design, and machine parameters. Spindle torque is a key variable in FSW since it is directly related to heat generation. The compiled data set permitted analysis of spindle torque from a wide variety of welding conditions. To compare heat generation from diverse welding procedures, the torque data and welding tool geometries were used to calculate the average contact shear stress during welding, which was analyzed against the average tool surface velocity for each procedure. The results gave insight into the effects of welding speed, rotational speed, and alloy on the average flow stress. The results suggested evidence of reaching a minimum flow stress at high tool surface velocities and a possible effect of initial workpiece temper on conditions during welding.

Keywords Joining · Aluminum · Heat generation

Introduction

FSW has been a topic of research for nearly three decades, and a sizable body of work has been published. Most research has been focused on specific materials, alloys, workpiece thickness, etc., limiting each scope simply as a practical consideration. One exception has been the work by researchers at the University of South Carolina which analyzed welds from a wide spectrum of welding conditions in aluminum alloys 2524, 7050, and others [1–3], all of which included spindle torque measurements. The present work sought to collect data from many published works to examine broad trends, restricting focus only to conventional FSW in aluminum alloys. These data, combined with unpublished procedures, gave insight into heat generation in friction stir welding.

K. Colligan (✉)
Concurrent Technologies Corporation, Johnstown, PA, USA
e-mail: colligak@ctc.com

Y. Hovanski et al. (eds.), *Friction Stir Welding and Processing XII*,
The Minerals, Metals & Materials Series,
https://doi.org/10.1007/978-3-031-22661-8_7

Heat generation is an important parameter in any welding process, as it relates to the effect of welding on the materials being joined, to the productivity of the process, and to the energy cost in commercial welding operations, as examples. It is also an indicator of the conditions present during welding. FSW is a fully coupled thermomechanical process, meaning that the material response develops a self-referential balance: heat dissipation preconditions the workpiece material, setting the stage for heat generation in the weld, which then dissipates into the surrounding material. This balance is responsible for the inherent stability of the FSW process. This also means that the heat input is determined by the interaction between the welding tool and the workpiece and is a response from the process, not a prescribed input to the process. As such, heat generation is an important indicator of what is happening during welding, with the proper analysis.

Researchers have developed various approaches to expressing heat generation in FSW, often as input to models for predicting temperature distribution during welding. The earliest approaches represented the heat as coming from the shoulder only, assuming a circular heat source on the plate surface to calculate heat input from an assumed friction coefficient or based on the workpiece shear flow stress, acting on the shoulder area [4–8]. Analysis of the contact conditions between the tool and matrix followed [9–13], leading to the conclusion that heat generation could be best represented as being based on the shear flow stress acting on all tool surfaces. In 2002, an important analytical step was made by relating the spindle torque to the shear flow stress of the matrix, assuming a uniform average flow stress distributed over the tool surfaces [9, 10].

A detailed analysis was developed by Schmidt et al. [11] to rationalize the tool/matrix contact shear stress value based on different categories of contact conditions, including sticking, sliding, and mixed sticking/sliding. In the case of sliding, the defined condition was one of the relative motion between the tool and the workpiece with the workpiece not sufficiently stressed to cause deformation. The case of sticking was defined such that there was seizure between the tool and the workpiece, resulting in matrix deformation at the local velocity of the tool at the contact interface, with decaying velocity moving away from the interface. Finally, the case of mixed sticking/sliding was defined such that the matrix material is deforming at the contact interface at a speed that was less than the local tool velocity and was thus sliding relative to the tool but was deforming, nonetheless. In these latter two cases, the contact stress was rationalized to be equal to the shear flow stress of the matrix—the difference between the two cases being the velocity of the matrix at the contact interface.

According to Schmidt, the contact conditions in FSW of aluminum generate plastic deformation over most of the tool surfaces, except for a small region at the periphery of the shoulder, where frictional heating generates very close to the same heat as the sticking condition. These conclusions confirm the assumption made in the input torque model approach that the spindle torque results from contact shear flow stress that is uniformly distributed over the tool. This is the approach taken for analysis of spindle torque and welding tool geometry data from the data set.

Data Set

Data were collected from published and unpublished sources [13]. All welding procedures represented conditions that produced void free welds, which may or may not have been optimized to any criteria. All procedures were for aluminum alloys welded using conventional, single-sided FSW. A total of 170 procedures, representing 20 different alloys in various tempers, were included, 134 of which included spindle torque data. The complete data set is available for download, with references included [14].

Analytic Approach

Spindle Torque Analysis

The shear yield stress is calculated by dividing the measured spindle torque by the surface integral of the radius,

$$\tau_{\text{yield}} = \frac{T}{\int_s r dA} = \frac{T}{\text{geo}} \tag{1}$$

where T is the spindle torque and S is the total surface of the tool, including the shoulder, the probe sides and the probe end. The surface integral is referred to as "geo" since it comprises the geometry of the welding tool. To derive geo, an assumption was made of a concave or flat shoulder, a frustrum or cylindrical probe, and a flat probe tip. The incremental areas for each region of the tool surface are

$$\begin{aligned} \text{Shoulder incremental area} &= \frac{r}{\cos\alpha} dr\, d\theta \\ \text{Probe side incremental area} &= \frac{r}{\cos\beta} dz\, d\theta \\ \text{Probe tip incremental area} &= r\, dr\, d\theta \end{aligned} \tag{2}$$

where
α = shoulder concavity angle
β = frustum probe half angle
S = shoulder radius
P_R = probe root radius
P_T = probe tip radius
r = dimension in radial direction
z = dimension along probe axis
t = workpiece thickness, or probe length

θ = anglular position

The radius dimension to the probe side is related to the coordinate along the probe axis, z, using the expression $r = (P_R - z\tan\beta)$. Then, noting that $dA = rdrd\theta$,

$$\begin{aligned}\text{geo} &= \int_s r dA \\ &= \int_0^{2\pi}\int_{P_R}^{S} \frac{r^2}{\cos\alpha} drd\theta \\ &+ \frac{1}{\cos\beta}\int_0^{2\pi}\int_0^{t}(P_R - z\tan\beta)^2 dzd\theta \\ &+ \int_0^{2\pi}\int_0^{P_T} r^2 drd\theta \end{aligned} \tag{3}$$

For cylindrical probes, P_T is set equal to P_R and β is set to zero, for flat shoulders, α is set to zero. Performing the integrations yields,

$$\text{geo} = \frac{2\pi}{3}\left[\frac{S^3 - P_r^3}{\cos\alpha} + \frac{t\left((P_R + P_T)^2 - P_R P_T\right)}{\cos\beta} + P_T^3\right] \tag{4}$$

Combining Eqs. (1) and (4),

$$\tau_{\text{yield}} = \frac{T}{\frac{2\pi}{3}\left[\frac{S^3 - P_r^3}{\cos\alpha} + \frac{t((P_R+P_T)^2 - P_R P_T)}{\cos\beta} + P_T^3\right]} \tag{5}$$

Finally, the normal flow stress is calculated from the shear flow stress using the von Mises yield criterion,

$$\sigma_{\text{yield}} = \tau_{\text{yield}}\sqrt{3}. \tag{6}$$

Equations (5) and (6) were used to calculate the flowtress for each welding procedure in the data set that included torque data, enabling study of flow stress variation as a function of average surface velocity for many different conditions, as will be presented in the following section.

Surface Velocity Analysis

Representation of the surface velocity has taken various forms in the literature, including surface velocities developed from the radius of the probe or the shoulder. Long et al. [15] examined spindle torque as a function of tool rotation speed, which sufficed for evaluation of torque from a single welding tool design. Since the present

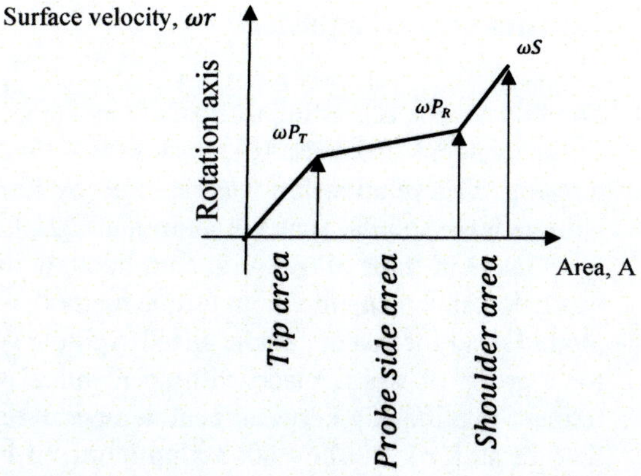

Fig. 1 Surface velocity as a function of area

study regarded torque measurements from a wide variety of welding tool geometries, it was necessary to develop a basis for comparison that comprised the essential tool features.

In the present study, an area-based average surface velocity was expressed as the velocity of an increment of area, integrated over the area, and divided by the total area. Imagine a function of surface velocity plotted against surface area, shown in Fig. 1.

The average of this function is given by

$$\text{Average surface velocity} = \omega\bar{r} = \frac{\omega \int_s r dA}{\text{total surface area}} \tag{7}$$

The integral in the numerator is equal to *geo*, from Eq. (4), so,

$$\omega\bar{r} = \omega\frac{\text{geo}}{\text{total area}} \tag{8}$$

resulting in

$$\omega\bar{r} = \omega\left[\frac{\frac{2}{3}\left[\frac{S^3-P_R^3}{\cos\alpha} + \frac{t\left((P_R+P_T)^2-P_RP_T\right)}{\cos\beta} + P_T^3\right]}{\frac{(S+P_R)(S-P_R)}{\cos\alpha} + \frac{t(P_R+P_T)}{\cos\beta} + P_T^2}\right] \tag{9}$$

Results and Discussion

The flow stress, according to Eqs. (5) and (6), was plotted as a function of average surface velocity from Eq. (9), as shown in Fig. 2. The data suggest a bilinear relationship. This relationship was reported by Long et al. [15], based on welds made with variable spindle speed in aluminum 2219, 5083, and 7050. In that case, welds were made in three alloys with continuously increasing spindle speed. Flow stress was calculated using the input torque method, as was done in the present study, then plotted as a function of spindle speed. Spindle speed is equivalent to surface velocity for a group of welds made with an identical welding tool geometry. Similarly, a bilinear relationship between peak temperature and spindle speed was reported by Sato et al. [16], in alloy 6063 aluminum. In both cases, a grain size plateau was reported at about the same rotational speed as a distinct flattening of the torque curve and the maximum temperature curve, respectively. Further increase in rotation speed gave diminished increase in grain size and peak temperature, with diminished decrease in torque. In these studies, the plateau in grain size was concluded to be an artifact of having reached a plateau of peak temperature since the grain growth was consistent with static recrystallization from a peak temperature. The results presented in Fig. 2 suggest that this phenomenon may be present in many aluminum alloys during FSW. Note that two welding procedures in 7075 aluminum were excluded from the least squares calculation for reasons given below.

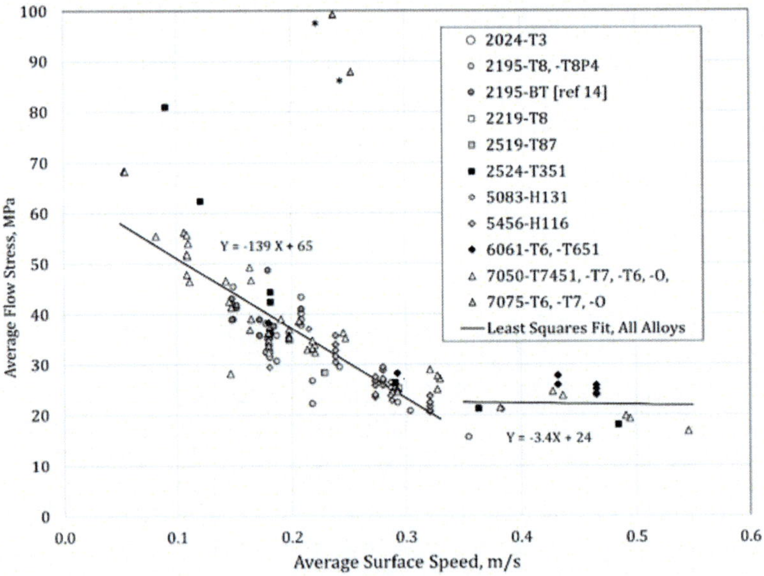

Fig. 2 Flow stress as a function of average surface speed, all alloys. *Indicates two procedures excluded from least squares fit

Details from specific alloys provide further insight. Average flow stress data for Al alloy 5456-H116, 5083-H131, and 5083-O plates are presented as a function of average surface velocity in Fig. 3, using spindle torque data from unpublished results by the author from Concurrent Technologies Corporation (aluminum 5456 and 5083), published results by the author [17] and published results from the University of South Carolina [2]. The aluminum 5456 welds were made as part of a study that produced welds with pre-rotation defects [18]. Resolution of the defects involved reducing the shoulder diameter, from about 35.6 to 30.5 mm. Five travel speeds and two rotational speeds were used for each tool. As can be seen in the figure, travel speed had the effect of increasing the flow stress, while increasing the spindle speed had the effect of decreasing the flow stress. These effects have been noted elsewhere [2, 13, 17]. A similar trend was discernable in the 7050 data, presented below. It is also notable that the welds in 5083-O made at low surface velocity had lower flow stress than the other 5XXX alloy welds, suggesting a lingering effect of the initial material condition, which was less pronounced at higher surface velocity and travel speed. Procedures in 7050 and 7075-O temper did not exhibit this temper effect. The range of average surface velocity in the test data was insufficient to demonstrate a certain agreement with heat-treatable alloys at high surface velocity.

Flow stress data from welds made in 2524-T351 are presented in Fig. 4. Here, there was close agreement with a bi-linear characterization. It is interesting that the slopes

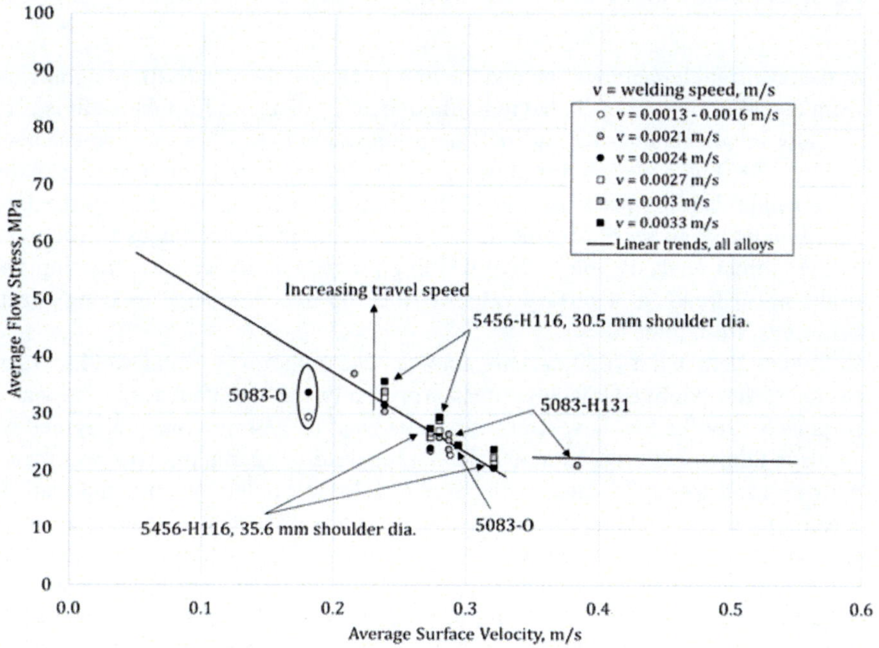

Fig. 3 Flow stress for 5456-H116 and 5083-O, -H131

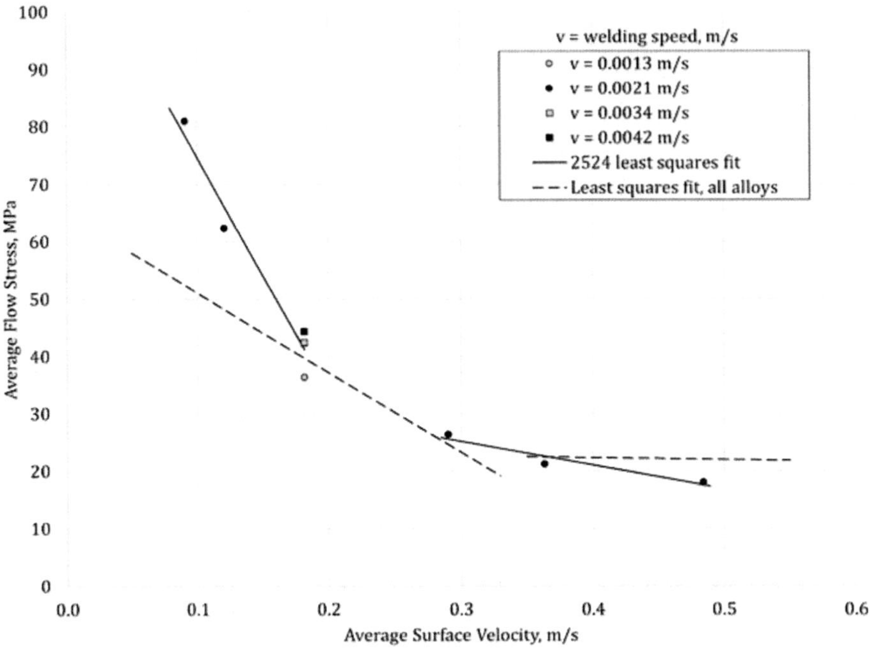

Fig. 4 Flow stress for 2524-T351

of the bi-linear characterization and the intersection of the trends differ significantly from the all-alloy average behavior, although no explanation for this result is given.

Flow stress data for Al alloy 7050 are presented in Fig. 5. The current data set for Al alloy 7050 was extracted from work at the University of South Carolina, except for a single welding procedure in 39.4 thickness material, indicated in the figure [from unpublished results by the author at Concurrent Technologies Corporation]. In the current analysis, alloy 7050 welding procedures indicated a crossing point in the linear trends at a surface velocity of 0.305 m/s, compared to Long's result, 0.224 m/s, from three alloys.

Two outliers in the 7050 data are notable. One outlier was from a weld in 32 mm thickness that exhibited low flow stress. Apparently, the high thickness alone was not responsible for the low flow stress, since the weld in 39.4 mm material agreed with welds in thinner material. It is notable that the weld in 32 mm material was made at a lower travel speed, 0.9 mm/s, compared to 1.3 mm/s in the 39.4 mm material. The lower flow stress would be consistent with lower travel speed, as discussed above, but the size of the effect is greater than observed in thinner materials. A second outlier came from a pair of welds made in 7075-T6 material at very high travel speed, 8.5 mm/s. The size of the effect of travel speed was much higher than that observed in 5456 aluminum, likely because the travel speed was so much higher. These high travel speed welds may represent conditions where material flow is contributed to by

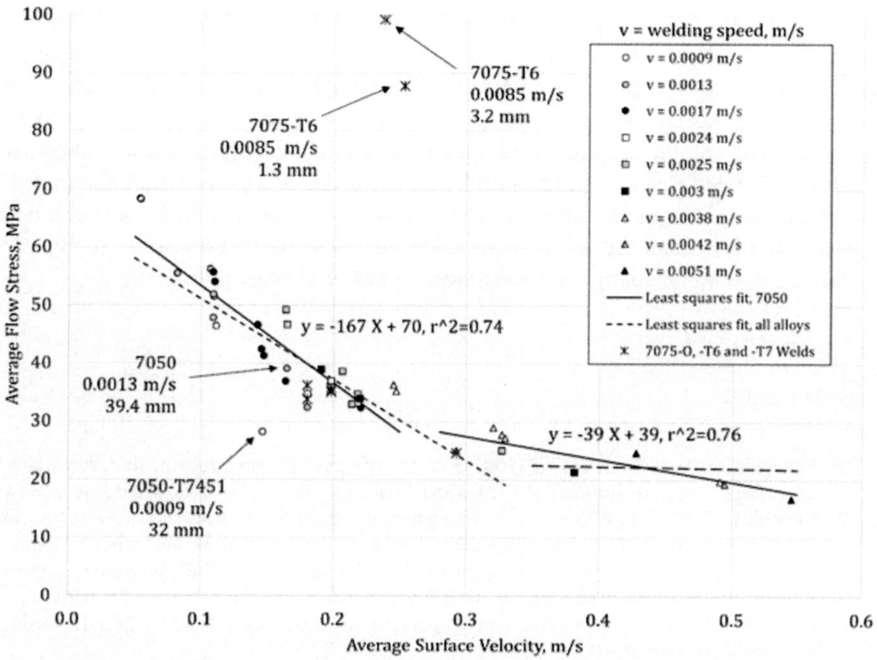

Fig. 5 Flow stress for 7050-O, -T7 and -T7451 with 7075-O, -T6 and -T7, for comparison

forcible extrusion around the probe. Further research into flow stress variation with travel speed is needed.

It was tempting to surmise from the general trend of the data in Fig. 2 that there may have been a continuous inverse relationship between flow stress and surface velocity. However, examination of the details from Al alloys such as 2524 and 7050, combined with published results from Long and Sato mentioned above, suggest that some step-change occurs, producing a discontinuous relationship. This step change may simply be due to approaching the melting point of the matrix. Further research is needed to explore this in detail for different aluminum alloys.

Conclusions

Welding procedures from published and unpublished sources were collected and analyzed to extract useful trends in the data. Analysis of torque from the data set gave insight into the response of various aluminum alloys to the conditions of welding. When plotted against average tool surface velocity, a bi-linear relationship was observed. Average flow stress values of 20 MPa to 60 MPa were most common. The results confirm observations from other researchers, with a more extensive data set than existed previously, and suggest further research.

The results suggest control strategies for closed-loop welding control systems. For example, it may be reasonable for a controller to limit the rotational speed to a level where an average surface velocity of 0.3 m/s is reached, since further increase in rotational speed yields limited additional heat generation for the travel speeds studied. Further, the welding controller may command a welding speed increase or rotational speed decrease if the spindle torque drops to a point where the minimum flow stress for a specific alloy is reached, generally around 20 MPa. Such a drop in torque may result from a change in thermal boundary conditions, such as from welding past some change in workpiece geometry, for example.

References

1. Yan J, Suttton MA, Reynolds AP (2005) Process-structure-property relationships for nugget and heat affected zone regions of AA2524-T351 friction stir welds. Sci Tech Weld Join, December
2. Reynolds AP, Tang W (2001) Alloy, tool geometry, and process parameter effects on friction stir weld energies and resultant FSW joint properties. In: Jata KV, Mahoney MW, Mishra RS, Semiatin SL, Field DP (eds) Friction stir welding and processing. TMS, Warrendale, pp 15–23
3. Reynolds A, Tang W, Khandkar Z, Khan JA, Lindner K (2005) Relationships between weld parameters, hardness distribution and temperature history in alloy 7050 friction stir welds. Sci Technol Weld Joining 10(2):190–199
4. Russell MJ, Shercliff HR (1998) Analytical modelling of friction stir welding. In: Proceedings INALCO '98: 7th International Conference: Joints in Aluminum, Vol 2, pp185–217, Cambridge, UK, 16 April
5. Frigaard O, Grong O, Midling OT (1998) Modelling of heat flow phenomena in friction stir welding of aluminum alloys. In: Proceedings INALCO '98: 7th International Conference: Joints in Aluminum, Vol 2, pp 185–217, Cambridge, UK, 16 April
6. McClure JC, Tang W, Murr LE, Guo X, Feng Z, Gould JE (1998) Thermal model in friction stir welding. In: 5th International Trends in Welding Research Conference Proceedings, 1–5 June, Pine Mountain, GA, ASM Int'l, pp 590–596
7. Russell MJ, Shercliff HR (1999) Analytical modelling of microstructure development in friction stir welding. In: 1st International Symposium on Friction Stir Welding, Thousand Oaks, CA, 14–16 June
8. Chao YJ, Qi X (1999) Heat transfer and thermo-mechanical analysis of friction stir joining of AA6061-T6 plates. In: 1st International Symposium on Friction Stir Welding, Thousand Oaks, CA, 14–16 June
9. Khandkar MZH, Khan JA, Reynolds AP (2003) Input torque based thermal model of friction stir welding. In: 6th International Trends in Welding Research Conference Proceedings, 15–19 April, 2002, Pine Mountain, GA, ASM International, pp 218–223
10. Khandkar MZH, Khan JA, Reynolds AP (2003) Prediction of temperature distribution and thermal history during friction stir welding: input torque based model. Sci Technol Weld Joining 8(3):165–174
11. Schmidt H, Hattel J, Wert J (2004) An analytic model for the heat generation in friction stir welding. Modell Simul Mater Sci Eng 12:143–157
12. Schmidt H, Hattel J (2005) A local model for the thermomechanical conditions in friction stir welding. Modelling Simul Mater Sci Eng 13:77–93
13. Colligan K (2022) Analysis of data from friction stir welds in aluminum. In: Paper presented at the 2022 International Symposium on Friction Stir Welding and Processing, Luneburg, Germany, 28–30 September
14. https://www.ctc.com/friction-stir-welding

15. Long T, Tang W, Reynolds AP (2007) Process response parameter relationships in aluminum alloy friction stir welds. Sci Tech Weld Join 12(4):311–317
16. Sato YS, Urata M, Kokawa H (2002) Parameters controlling microstructure and hardness during friction stir welding of precipitation-hardenable aluminum alloy 6063. Metall Mater Trans A 33A:625–635
17. Colligan KJ, Xu J, Pickens JR (2003) Welding tool and process parameter effects in friction stir welding of aluminum alloys. In: Jata KV, Mahoney MW, Mishra RS, Lienert T (eds) Friction stir welding and processing II. TMS, Warrendale, pp 181–190
18. Colligan K (2010) Description of a pre-rotation defect in friction stir welding of 5456 aluminum. In: 8th International Symposium on Friction Stir Welding, Timmendorfer Strand, Germany, May 18–20

Part IV
FSW of High Melting Temperature Materials

Assessing Manufacturability of the Oxide Dispersion Strengthened (ODS) 14YWT Alloy Fuel Cladding Tube Using SolidStir™ Technology

Shubhrodev Bhowmik, Pranshul Varshney, Osman El Atwani, Stuart A. Maloy, Kumar Kandasamy, and Nilesh Kumar

Abstract The 14YWT (Fe–14Cr–3W–0.4Ti–0.3Y_2O_3 (wt.%)) alloy is a potential candidate material for fabricating fuel cladding tubes and other structural components to be used in next-generation advanced nuclear reactors. Due to inherent limitations of current processing routes, this research explored a new variant of friction stir processing, commercially referred to as SolidStir™ Extrusion (patent pending) technology from Enabled Engineering to consolidate and extrude 14YWT powder into a fuel cladding tube. Several single-layer and multiple-layer consolidation experiments were carried out in an Inconel 625 die to determine the suitable processing parameters for SolidStir™ extrusion of the 14YWT alloy cladding tube. With these processing parameters, a 16–18 mm long tube of the 14YWT alloy with no visible defects was successfully extruded using this novel technology. Microstructural characterization results obtained using optical microscopy, scanning electron microscopy, energy dispersive spectroscopy, electron backscatter diffraction, and transmission electron microscopy will be presented.

Keywords ODS 14YWT alloy · SolidStir · Solid-state processing · Fuel cladding tube

---

S. Bhowmik · P. Varshney · N. Kumar (✉)
Department of Metallurgical and Materials Engineering, The University of Alabama, Tuscaloosa, AL 35487, USA
e-mail: nkumar@eng.ua.edu

O. E. Atwani
Los Alamos National Lab, Los Alamos, NM 87545, USA

S. A. Maloy
Pacific Northwest National Lab, Richland, WA 99354, USA

K. Kandasamy
Enabled Engineering, Blacksburg, VA 24060, USA

Y. Hovanski et al. (eds.), *Friction Stir Welding and Processing XII*,
The Minerals, Metals & Materials Series,
https://doi.org/10.1007/978-3-031-22661-8_8

Introduction

The designs of next-generation advanced nuclear reactors (i.e., Gen IV) have to incorporate high-performance structural materials that can sustain irradiation doses of more than 200 dpa while being subjected to high temperature (good creep rupture strength at 550–900 °C) for several years [1–4] for several structural components including fuel cladding tubes. Consequently, among many high-performance materials, oxide dispersion strengthened (ODS) ferritic steels are being explored as potential structural materials to produce fuel cladding tubes. In a ferritic matrix, these steels have a homogeneous dispersion of Y–Ti-O oxide nanoparticles, also referred to as nanoclusters. The ferritic matrix resists radiation-induced swelling, while the oxide nanoparticles add to mechanical strength by pinning dislocations and grain boundaries at high temperatures, as well as reduce void swelling. Nanoparticle-matrix interfaces and grain boundaries operate as sinks for radiation-induced defects, reducing irradiation damage [4–7].

Currently, multiple-step extrusion is employed to manufacture ODS cladding tubes [4]. In this method, pre-alloyed ferritic matrix powders are mechanically alloyed with yttria particles in an attritor ball mill for 40 h. After that, powders are bottled, degassed under vacuum, sealed, and hot extruded at 1123 K (~850 °C). Following a Pilger mill cold roll, the cladding tubes are next heated to 1353 K (~1079 °C) for annealing. In accordance with the desired final dimension, this step (cold rolling + annealing) is performed 4 to 10 times. After the last cold rolling pass, the tubes are heat-treated at 1373 K (~1099 °C) for 1 h or 1423 K (~1149 °C) for 60 s to reduce dislocation density. Final manufacturing steps include polishing and sandblasting [4, 8]. While uniformly spreading the oxide nanoclusters in the ferritic matrix, this traditional production process creates cladding tubes with precise dimensions. Nonetheless, the manufacture of the tubes is time- and money-intensive due to such several-step, low output rate techniques. Moreover, the tubes may produce anisotropic grains and, as a result, may have anisotropic mechanical properties which is another effect of using such standard manufacturing techniques [4, 9].

To address these issues, this study developed a novel method commercially referred to as "SolidStir™ extrusion technology" for extruding a 14YWT tube using the principles of friction stir welding or processing (FSW/P) technology. SolidStir™ is a patent pending and proprietary technology of Enabled Engineering. Consolidated ball milled 14YWT powder was extruded via a die using a W–Re tool. The FSW/P processing parameters were determined by carrying out 14YWT powder consolidation in an Inconel 625 close-die experiments. The information on the microstructure of the consolidated powders was obtained using optical microscopy, scanning electron microscopy (SEM), electron backscatter diffraction (EBSD), and transmission electron microscopy (TEM) techniques. Using slight adjustment in optimized processing parameters obtained in closed-die experiments during tube extrusion, a defect-free tube of 14YWT alloy was successfully obtained.

Materials and Method

The 14YWT alloy powder, obtained by ball milling a mixture of gas-atomized Fe-14Cr-3 W-0.4Ti-0.2Y and FeO powder for 40 h, was supplied by Los Alamos National Laboratory (LANL). The particle size was determined by taking the average of at least 25 different particle sizes and then thresholding the images using ImageJ. The Bond Technologies, FSW machine from the RM (research machine) series was used. FSW parameter tuning required several experiments. They were in 3 steps. Step 1 involved drilling blind holes (9.99 mm diameter and 2.3 mm depth) in an Inconel 625 plate to make dies (Fig. 1). Then, the hole was filled with 14YWT powder and consolidated by the heat and strain generated from the rotation of the W–Re tool (Fig. 1). At this stage, the optimal settings for the tool rotation speed, forge velocity, dwell duration were determined to be 500 rpm (rotation per minute), 5 mm/s, and 10 s, respectively, after characterizing the microstructure. Following the adjusted parameters, more experiments were carried out on an Inconel 625 die in Step 2. These trials involved raising the blind hole depth to a value of 5 mm to get more volume of consolidation than Step 1, maintaining the tool in its previous state, and shielding the processing with argon (Ar) gas. Powder consolidation was used to make samples with both a single layer and multiple (3) layers. In order to determine the degree of consolidation present in the samples, microstructural analysis was performed. In Step 3, which involved tube extrusion, the tool form needed to be adjusted so that the powders could be extruded through the die. When manufacturing tubes with the SolidStir™ Technology (SolidStir-Ex1200Core), the optimal tool rotational rate of 500 rpm was employed. The setup for tube extrusion is shown in Fig. 1.

Low-magnification optical microscopy was used to capture macroscopic view of the 14YWT tube by a Keyence Digital Microscope. A scanning electron microscope (JEOL 700) was used to identify microstructural characteristics such as chemical

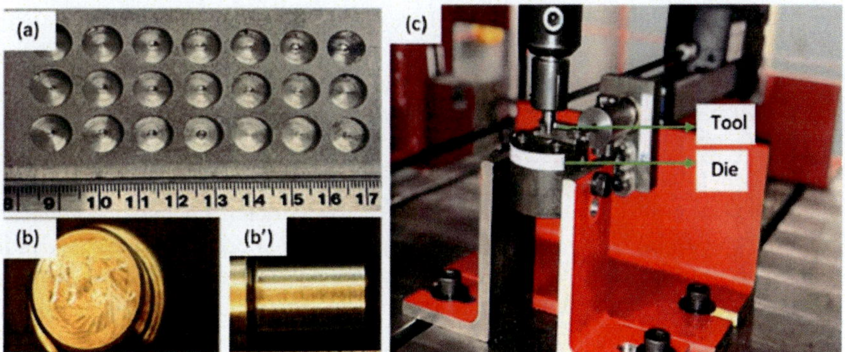

Fig. 1 Images showing an Inconel 625 die with blind holes (**a**) and a W–Re tool (**b**, **b′**) that was utilized for parameter selection trials; tube extrusion experimental setups (**c**) are also displayed

makeup, grain size, nature of grain boundaries, and texture. Depending on the grain size and feature that needed to be captured, EBSD (electron backscatter diffraction) was performed at magnifications ranging from 2,000 to 12,000. The EBSD results were post-processed by TSL OIM 8 and texture was analyzed by the MTEX toolbox of MATLAB. The SEM micrographs were captured in backscatter electron imaging (BEI) and secondary electron imaging (SEI) modes for consolidated powder samples and extruded tubes. The extruded tube was cut with a slow-speed diamond saw. The tube was sectioned, and the EBSD was performed on the longitudinal cross section while macrostructural images of the transverse cross section were taken by an SEM. The grinding and polishing of the samples were performed by using SiC papers up to 800 grits followed by 9 μm to 1 μm diamond polishing solution. 0.05 μm alumina and 0.02 μm colloidal silica were used for the final polishing stage. FEI Tecnai F20 was used to investigate precipitate shape and composition through energy dispersive spectroscopy (EDS) in TEM. The TEM sample was prepared via focused ion milling.

Results and Discussion

Figures 2a, a′ show a macroscopic view of the transverse cross section of the consolidated powder achieved in Step 2 obtained using SEM in SEI (Fig. 2a) and BEI (Fig. 2a′) modes. A highly consolidated region was observed at the top of the consolidated sample probably due to high temperature and forging pressure existing in the region just beneath the tool tip. No segregation was observed from BEI imaging proving the homogeneous mixing and consolidation of the powder. Figures 2b, b′ include the macrostructure taken in both BEI and SEI modes in an SEM for a multi-layered consolidated sample. There was a significant amount of bonding between layers despite showing some voids at the interface. This was an indication that longer tubes can be continuously extruded, where during extrusion process, multiple layers of the tube are microscopically welded together.

The EBSD scans at different layers revealed a range of grain sizes (Fig. 3). Layers 1 (top), 2 (middle), and 3 (bottom) have respective grain sizes of 0.98 μm, 1.27 μm, and 0.50 μm. The phase used to index in EBSD was ferrite or body-centered cubic (BCC). Since the scan was conducted at a very high resolution or magnification, the number of grains for layer 2 cannot be precisely determined in accordance with ASTM standard E112-13. Nonetheless, a qualitative interpretation was possible.

In addition, the fraction of low-angle grain boundaries (LAGB) reduced from layer 1 to layer 2 (0.80–0.43). The larger grain size in layer 2 may be a result of grain growth due to heat input during the consolidation of the powders in layer 1. The reduction in LAGB was also indicative of recovery and grain growth in layer 2. According to the literature, this is a regular occurrence for additive manufacturing techniques in general specially for additive friction stir deposition (AFSD) which is also based on FSW/P principles [10]. It is believed that the grains might have grown in layer 3 as well because of layer 2 deposition over layer 3. However, the effect of

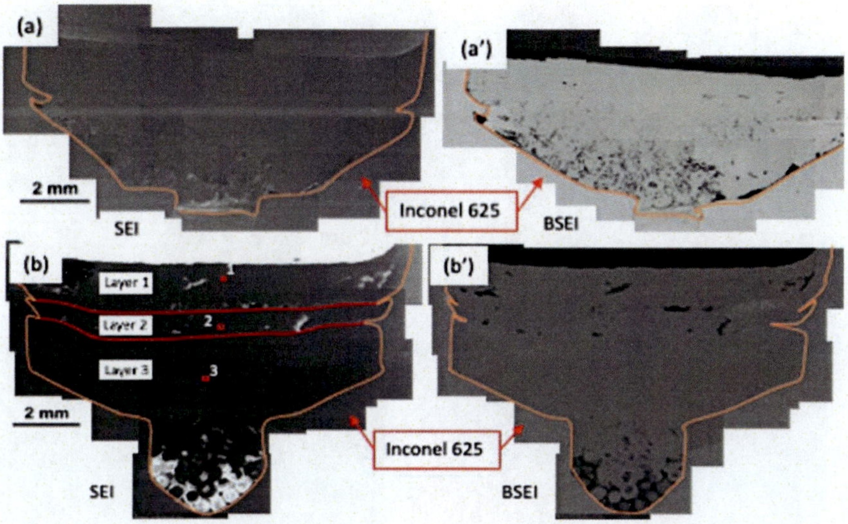

Fig. 2 SEM macrographs of the consolidated powder samples; (**a**, **a′**) single-layer consolidation and (**b**, **b′**) multilayer consolidation

deposition of layer 1 on the grain growth in layer 3 is expected to be low. Due to the effect of layer 2 on layer 3 and of layer 1 on layer 2, layer 2 and layer 3 are expected to have similar fraction of LAGB. However, it is not clear why there is difference in the fraction of LAGB between layer 2 and layer 3. It is quite possible that the EBSD scan size does not truly represent the actual microstructure of layer 2. A larger area EBSD scan of layer 2 will be carried out in future to clarify this discrepancy.

For proof-of-concept demonstration purpose, only 16–18 mm long tube was extruded from the 14YWT alloy. At macroscopic level, neither the inner nor the outer surface of the extruded tube displayed any obvious surface defects (Fig. 4a). The transverse cross section of the extruded tube was uniformly circular (Fig. 4b). The tube thickness was measured to be 500 μm. The tube of this length was extruded in roughly three minutes. According to the literature, the appearance of the tube was comparable to tubes created by other processing routes for fuel cladding applications [11, 12]. However, because this innovative extrusion method is conducted in a single step and at high rate, it is considerably faster than the conventional manufacturing process [8, 13].

The longitudinal section of the tube revealed elongated grains along the 45° of the ED (extrusion direction) using EBSD (Fig. 5a). Note that the phase utilized to index was ferrite or body-centered cubic (BCC). This may have occurred due to shearing during the extrusion process. The average grain size was determined to be 2.53 ± 1.26 μm, and the LAGB fraction was 0.30. The texture was analyzed by plotting orientation distribution function (ODF) map (Fig. 5b) and pole figures (Fig. 5c). It was found that the texture was not very strong, but there was a presence of weak γ-fiber, and the fiber itself had $(111)[2\bar{3}1]$, $(111)[1\bar{3}2]$ components. Although the

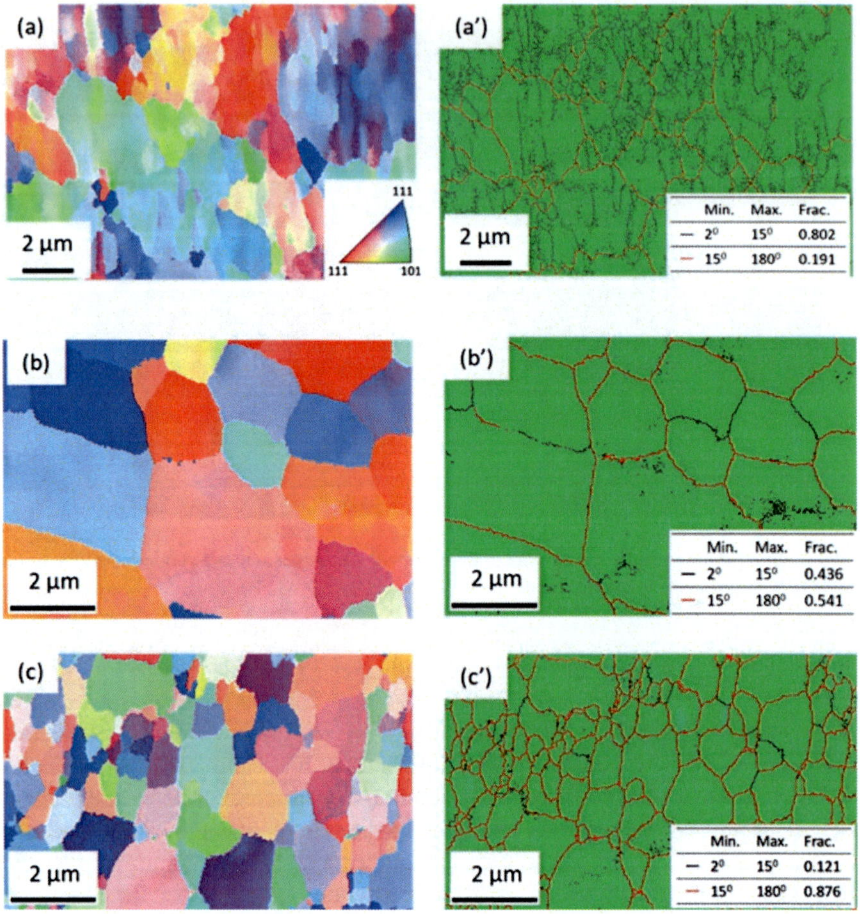

Fig. 3 IPF map with the presence of low-angle and high-angle boundaries obtained from EBSD at different layers of powder consolidated sample. **a**, **a′** layer 1, **b**, **b′** layer 2, **c**, **c′** layer 3. Forging direction (FD) is along the vertical axis of the map. The colors in the IPF maps (**a**, **b**, and **c**) represent the orientation of the FD sample axis with respect to the crystal lattice frame according to the coloring in the standard IPF triangles included as inset in **a**. Micrographs included in **a′**, **b′**, and **c′** show low-angle and high-angle grain boundaries

existence of α-fiber was also not continuous, many components were found including $(114)[1\bar{1}0]$, $(113)[1\bar{1}0]$, $(223)[1\bar{1}0]$, $(111)[1\bar{1}0]$. The existence of α and γ-fiber has been documented in the literature for several 14YWT tube processing procedures [14]. For more in-depth texture analysis, EBSD on a larger area will be performed to get a statistically significant number of grains.

The existence of precipitates or nanoclusters can be observed using BF-TEM (Fig. 6). The nano-sized precipitates were detected both at the grain boundaries (green arrow) and within the grains (red arrow). The EDS equipped in TEM was done to check the composition of the precipitates. The compositional map revealed that the

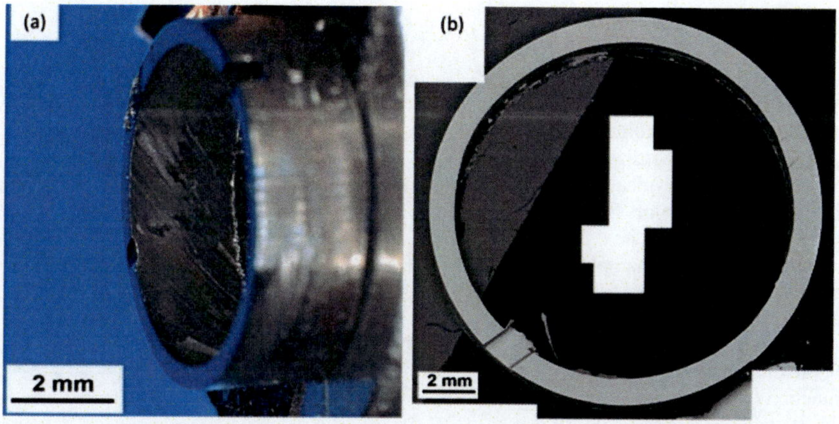

Fig. 4 Interior surface of the extruded tube (**a**) and the transverse cross section (**b**) showing uniform thickness of the wall of the extruded tube. The discontinuity present in the lower left corner of the tube cross section is due to machining mark left during sectioning process of the tube

precipitates were primarily composed of Y–Ti-O. In the future, high-resolution TEM (HRTEM) and atom probe tomography (APT) will be used to conduct a comprehensive investigation of the precipitates in order to collect more data on the interface properties and composition distribution.

Conclusion

In this study, a new processing method called SolidStir™ extrusion was used to manufacture fuel cladding tube of the 14YWT alloy. This technique was based on the principles of friction stir welding or processing. To find the best parameters for extrusion, a number of surrogate experiments were done. Micro- and macrostructural analysis showed that both single- and multi-layered samples had a high level of consolidation and that the layers of the multi-layered sample were well integrated. The EBSD showed that the grain sizes changed from the top to the bottom layers because more heat was added to the layers that were already consolidated. By using the optimized parameters, the 14YWT tube was made with a uniform circular cross section and no visible significant surface flaws. It was noticed that grains were elongated, and there were discontinuous γ and α fibers present. But the texture was not very strong. The TEM analysis showed that there were Y–Ti-O precipitates at both the grain boundaries and inside the grains. These are important for high-temperature creep and irradiation resistance in fuel cladding tubes. EBSD, TEM, and APT will be used to find out more about the texture and precipitates.

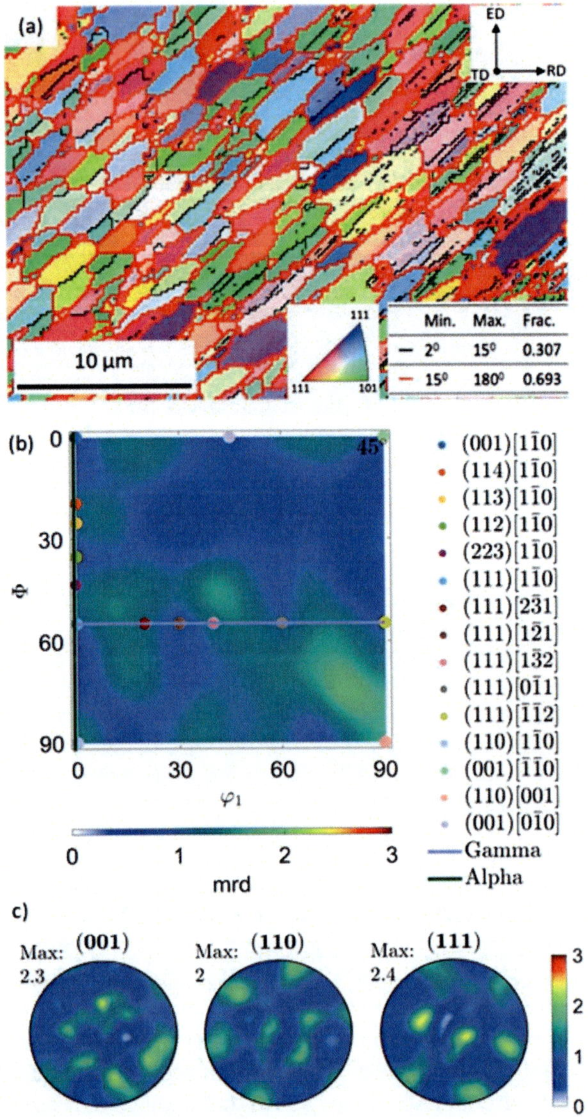

Fig. 5 IPF map of 14YWT tube displayed together with low-angle (black lines) and high-angle (red lines) grain boundaries on longitudinal cross section **a**. According to the coloring in the typical IPF triangles, the colors in the IPF maps showed the orientation of the ED sample axis with respect to the crystal lattice frame. Additionally, presented is the orientation distribution function (ODF) map at $\varphi_2 = 45^{\circ}$ **b** and pole figures (PF) **c** both of which indicate presence of weak texture. The sample frame for EBSD, ODF, and PF was the same

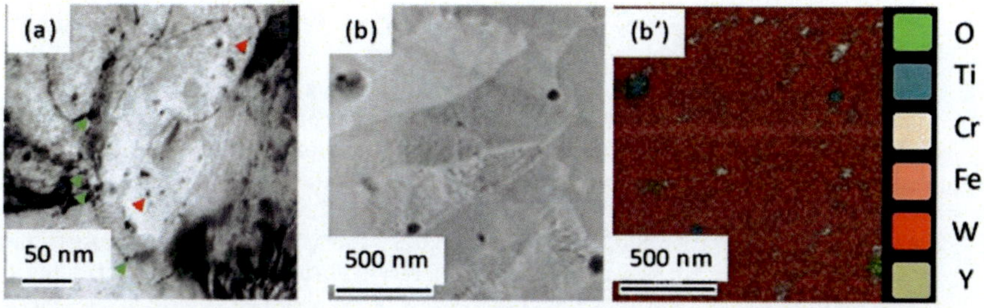

Fig. 6 Bright field (BF) TEM (**a**) image of the nanoclusters or precipices of the extruded 14YWT tube. The HADAF—STEM image (**b**) along with the EDS composition map (**b′**) are also shown to provide the composition of the precipitates

Acknowledgements This work was funded by DOE (award number: DE-SC0021794) through Enabled Engineering. We also acknowledge the use of microstructural characterization facility at Alabama Analytical Research Center at UA.

References

1. Murty K, Charit I (2008) Structural materials for Gen-IV nuclear reactors: Challenges and opportunities. J Nucl Mater 383(1–2):189–195
2. Kohyama A, Hishinuma A, Gelles D, Klueh R, Dietz W, Ehrlich K (1996) Low-activation ferritic and martensitic steels for fusion application. J Nucl Mater 233:138–147
3. Klueh R, Shingledecker J, Swindeman R, Hoelzer D (2005) Oxide dispersion-strengthened steels: a comparison of some commercial and experimental alloys. J Nucl Mater 341(2–3):103–114
4. Maier B, Lenling M, Yeom H, Johnson G, Maloy S, Sridharan K (2019) A novel approach for manufacturing oxide dispersion strengthened (ODS) steel cladding tubes using cold spray technology. Nucl Eng Technol 51(4):1069–1074
5. Wu Y, Ciston J, Kräemer S, Bailey N, Odette GR, Hosemann P (2016) The crystal structure, orientation relationships and interfaces of the nanoscale oxides in nanostructured ferritic alloys. Acta Mater 111:108–115
6. McClintock DA, Sokolov MA, Hoelzer DT, Nanstad RK (2009) Mechanical properties of irradiated ODS-EUROFER and nanocluster strengthened 14YWT. J Nucl Mater 392(2):353–359
7. Sagaradze V, Shalaev V, Arbuzov V, Goshchitskii B, Tian Y, Qun W, Jiguang S (2001) Radiation resistance and thermal creep of ODS ferritic steels. J Nucl Mater 295(2–3):265–272
8. Ukai S, Yoshitake T, Mizuta S, Matsudaira Y, Hagi S, Kobayashi T (1999) Preliminary tube manufacturing of oxide dispersion strengthened ferritic steels with recrystallized structure. J Nucl Sci Technol 36(8):710–712
9. Alamo A, Regle H, Pons G, Béchade JL (1992) Microstructure and textures of ODS ferritic alloys obtained by mechanical alloying. Trans Tech Publ, Materials Science Forum, pp 183–190
10. Williams M, Robinson T, Williamson C, Kinser R, Ashmore N, Allison P, Jordon J (2021) Elucidating the effect of additive friction stir deposition on the resulting microstructure and mechanical properties of magnesium alloy WE43. Metals 11(11):1739
11. Aydogan E, Pal S, Anderoglu O, Maloy S, Vogel S, Odette G, Lewandowski J, Hoelzer D, Anderson I, Rieken J (2016) Effect of tube processing methods on the texture and

grain boundary characteristics of 14YWT nanostructured ferritic alloys. Mater Sci Eng, A 661:222–232
12. Lenling M, Yeom H, Maier B, Johnson G, Dabney T, Graham J, Hosemann P, Hoelzer D, Maloy S, Sridharan K (2019) Manufacturing oxide dispersion-strengthened (ODS) steel fuel cladding tubes using the cold spray process. JOM 71(8):2868–2873
13. Vakhitova E, Sornin D, Barcelo F, François M (2017) Texture evolution in Oxide Dispersion Strengthened (ODS) steel tubes during pilgering process. J Nucl Mater 494:20–28
14. Aghamiri S, Sowa T, Ukai S, Oono N, Sakamoto K, Yamashita S (2020) Microstructure and texture evolution and ring-tensile properties of recrystallized FeCrAl ODS cladding tubes. Mater Sci Eng, A 771:138636

Effect of Welding Parameters on Microstructure and Mechanical Properties of Friction Stir Lap Welds of an Ultrahigh Strength Steel

Yutaka S. Sato, Shunsuke Mimura, Shun Tokita, Yusuke Yasuda, Akihiro Sato, and Satoshi Hirano

Abstract In this study, effect of welding parameters on microstructure and mechanical properties of an ultrahigh strength steel (UHSS) friction stir lap welded with Co-based alloy tool was examined. Friction stir lap welding of the UHSS was successfully done at various welding parameters. A decrease in rotational speed decreased fraction of martensite and hardness in the stir zone. Tensile shear load of the weld increased with decreasing the rotational speed. In the welds produced at the high rotational speed, crack was initiated from the tip of the lapped interface and then propagated into the stir zone having high fraction of martensite, possibly resulting from the stress concentration. On the other hand, the welds produced at the low rotational speed failed in the softest region of the heat-affected zone. This study showed that the reduced brittleness of the stir zone arising from the low rotational speed was an effective strategy for the superior mechanical properties of the lap welds of the UHSS.

Author Contribution: Shunsuke Mimura, Shun Tokita, Yusuke Yasuda, Akihiro Sato, and Satoshi Hirano contributed equally to this work.

Y. S. Sato (✉) · S. Mimura · S. Tokita
Department of Materials Processing, Tohoku University, 6-6-02 Aramaki-Aza-Aoba, Aoba-Ku, Sendai 980-8579, Japan
e-mail: ytksato@material.tohoku.ac.jp

S. Mimura
e-mail: shunsuke.mimura.p8@dc.tohoku.ac.jp

S. Tokita
e-mail: shun.tokita.c4@tohoku.ac.jp

Y. Yasuda · A. Sato · S. Hirano
Research & Development Group, Hitachi, Ltd, 7-1-1 Omika-Cho, Hitachi, Ibaraki 319-1292, Japan
e-mail: yusuke.yasuda.jf@hitachi.com

A. Sato
e-mail: akihiro.sato.xt@hitachi.com

S. Hirano
e-mail: s_hirano@iot.ac.jp

Y. Hovanski et al. (eds.), *Friction Stir Welding and Processing XII*,
The Minerals, Metals & Materials Series,
https://doi.org/10.1007/978-3-031-22661-8_9

Keywords Friction stir lap welding · Ultrahigh strength steel · Microstructure · Mechanical properties

Introduction

Various kinds of ultrahigh strength steels (UHSS) have already been used in the automotive industry for both the reduction of CO_2 emission and the collision safety. In general, thin UHSS sheets in the lap configuration are spot welded, but lap linear welding of them is expectedly effective for the highly rigid automotive bodies. Many papers on friction stir spot welding (FSSW) of UHSS [1–3] have been reported. Additionally, many studies on friction stir lap welding (FSLW) of dissimilar metals [4–7] have also been done. However, there are only a few studies on FSLW of steels. Ghosh et al. [8] carried out FSLW of high strength martensitic steel sheets with additional cooling at the different travel speeds and then examined the microstructure and mechanical properties of the weld. The weld failed along the heat-affected zone (HAZ) with the maximum drop of microhardness during tensile lap-shear testing, and thus, it was concluded that characteristics of the fracture location was key for the joint efficiency. This study showed the useful knowledge on FSLW of high strength steel sheets, but effect of rotational speed, strongly affecting the maximum temperature during friction stir welding [9], on microstructure and mechanical properties of FSLWed UHSS has hardly been clarified. In this study, two sheets of an UHSS were welded by FSLW with a Co-based alloy tool, and effect of welding parameters on microstructure and mechanical properties of the weld attempted to be clarified.

Experimental Procedures

UHSS sheet used in this study was a commercial 1Cr-0.25Mo steel subjected to quenching and tempering, having a tensile strength of about 1.5 GPa and an elongation higher than 7%. The sheet dimension was 74 × 220 × 1.4 mm. Two sheets were lapped (the lapped width was 22 mm), and then, FSLW was performed with a Co-based alloy tool [10]. Two lap welds were made on a set of two sheets with the different tool rotations, as schematically shown in Fig. 1. Tool consisted of the shoulder with the diameter of 15 mm and the conical probe with the length of 1.6 mm. In this study, the constant advance per revolution (APR), which was defined as APR = travel speed/rotational speed [11], of 0.5 mm was used by varying the rotational speed from 100 to 300 rpm and the travel speed from 50 to 150 mm/min.

Following FSLW, the lap welds were cut perpendicular to the welding direction by electrical discharge machine. The cross section was grounded and then finally polished with a diamond paste of 1 μm. Etching with a nital solution (5% nitric acid + 95% ethanol) was employed after polishing. The microstructure was examined by optical microscopy and scanning electron microscopy (SEM).

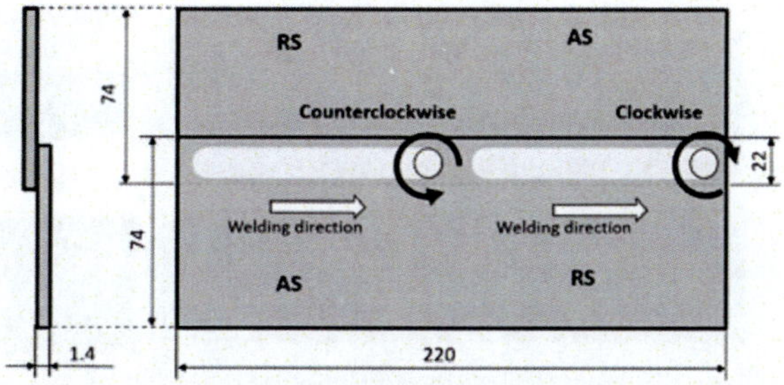

Fig. 1 Schematic drawing of FSLW

Mechanical properties of the weld were examined by Vickers hardness test and tensile shear strength test. Vickers hardness profile was measured at a load of 9.8 N along the midsection of the upper sheet on the cross section. Three samples of tensile shear strength test, 10 mm in width, were cut perpendicular to the welding direction. Tensile shear strength test was performed at a crosshead speed of 2.0 mm/min with an Instron-type tensile testing machine.

Results and Discussion

FSLW of the UHSS sheets was successfully done at all welding parameters used in this study. Cross-sectional overviews of the welds were shown in Fig. 2. The basin-shaped stir zones with no defects were obtained at all the welding parameters. The depth of the stir zone increased with increasing the rotational speed. The different contrast of basin-shaped stir zone arises from solid-state phase transformation during welding, i.e., the higher rotational speed causes the higher maximum temperature, resulting in the larger and deeper region where phase transformation occurred [12]. Fragmentation of the lapped interface in the stir zone was partly insufficient at 100 rm, while the lapped interfaces were well fragmented in the stir zone at the rotational speed higher than 150 rpm. The tip of the lapped interface was bended upward, i.e., the hook formed [13], at the advancing side, and the hook height slightly increased with increasing the rotational speed. On the other hand, the hook was hardly detected at the retreating side.

Vickers hardness profiles of the welds are presented in Fig. 3. The hardness of the base material was about 450 HV. The hardness was reduced toward the weld center, and the maximum reduction of the hardness was found in the HAZ proximal to the stir zone. The hardness of the softest HAZ was about 300 HV, hardly affected by the rotational speed. The stir zone showed the hardness between 380 and 500 HV, and the hardness value increased with increasing the rotational speed. The similar hardness profiles have been reported in previous papers on FSW of high strength steels [14, 15].

SEM images of the base material and the stir zone centers are presented in Fig. 4. The base material consists of the tempered martensite. The stir zone was composed of

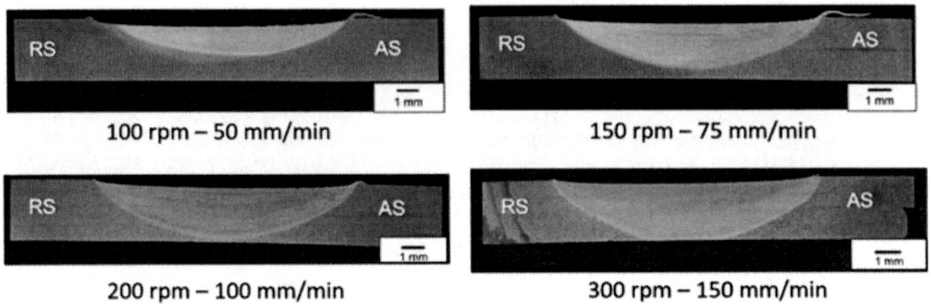

Fig. 2 Cross-sectional overviews of the lap welds

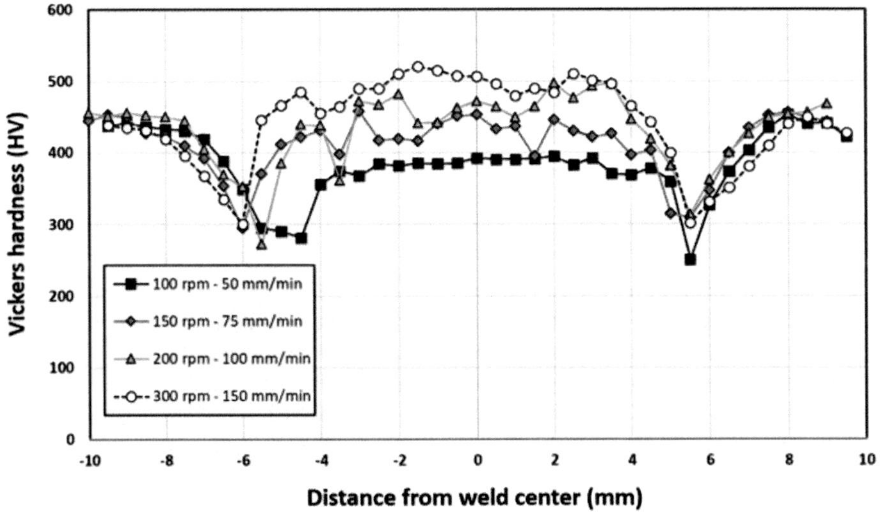

Fig. 3 Vickers hardness profiles of the lap welds

quenched martensite (dark phase) with ferrite (bright phase), and fraction of quenched martensite, which was quantified by point counting for SEM image, increased with increasing the rotational speed, as shown in Table 1. This is a reason for the higher hardness in the stir zone produced at the higher rotational speed. Formation of the quenched martensite in the stir zone arises from quenching of austenite stable at high temperatures during welding. Since hardenability of 1Cr-0.25Mo steel is high, the austenite would be entirely transformed into martensite during FSLW. The A_1 and A_3 temperatures of 1Cr-0.25Mo steel were calculated to 730˚C and 788˚C by Therm-Calc, respectively. From fraction of quenched martensite, the maximum temperatures during FSLW could be estimated (Table 1) because fraction of quenched martensite should change from 0 to 100% at the maximum temperatures from the A_1 to A_3 points. In the softest HAZ, large size of carbides in the tempered martensite was

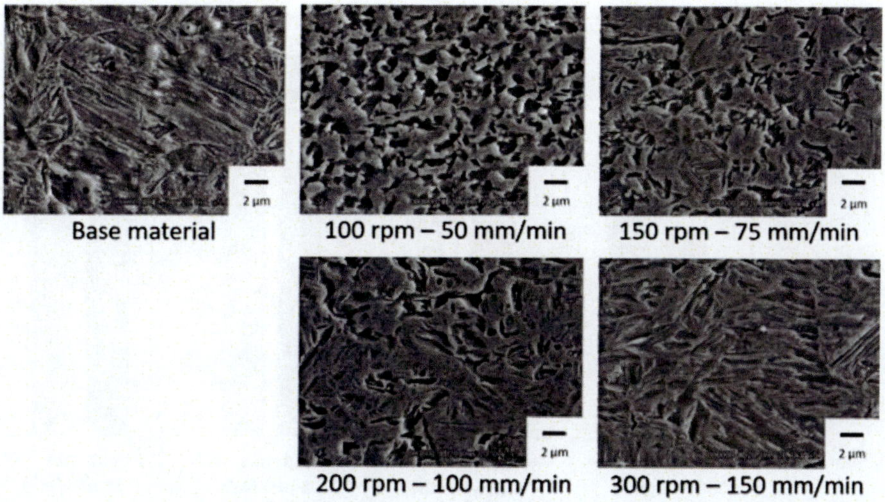

Fig. 4 SEM images of the base material and stir zone centers of the lap welds

Table 1 Area fraction of quenched martensite and estimated maximum temperature during FSLW

	100 rpm – 50 mm/s	150 rpm – 75 mm/s	200 rpm – 100 mm/s	300 rpm – 150 mm/s
Fraction of quenched martensite (%)	50	75	90	100
Estimated maximum temperature (°C)	740	760	780	> 788

observed. This means that the further tempering of the tempered martensite (of the base material) occurred in this region by thermal effect of FSLW.

Average tensile shear loads of the welds are summarized in Fig. 5. Tensile shear test showed that the lap welds produced at the lower rotational speed exhibited the higher tensile shear load. The load was maximum in the weld produced at 150 rpm, beyond which it decreased with increasing the rotational speed. The low tensile shear load at 100 rpm would arise from partly insufficient fragmentation of the lapped interface. Effect of rotational direction on tensile shear load was negligible at the low rotational speed.

To clarify the effect of rotational speed on tensile shear load, the crack initiation and propagation were specified on the cross sections of the fractured welds. SEM images of the crack initiation sites of the welds produced at 150 and 200 rpm, which were expressed as "low rotational speed" and "high rotational speed," respectively, are given in Fig. 6. In the welds produced at the high rotational speed, crack was initiated from the tip of the lapped interface (hook) and then propagated into the stir zone having high fraction of martensite. Reduction of the cross-sectional area during tensile testing was negligible, i.e., plastic deformation hardly occurred in these welds. Since the tip of the lapped interface would be an origin of the large

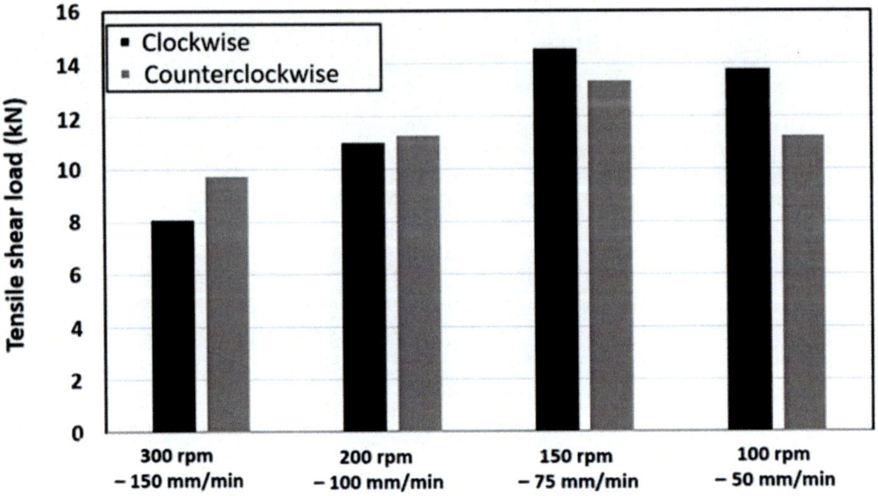

Fig. 5 Tensile shear load of the lap welds

stress concentration, it was implied that the brittle fracture occurred in the welds produced at the high rotational speed. In the lap welds produced at the low rotational speed, on the other hand, crack was initiated from the HAZ proximal to the stir zone and then propagated along the softest HAZ, arising from the reduced brittleness of the stir zone. This study showed that the low rotational speed was effective to produce the lap welds with the superior mechanical properties in the UHSS.

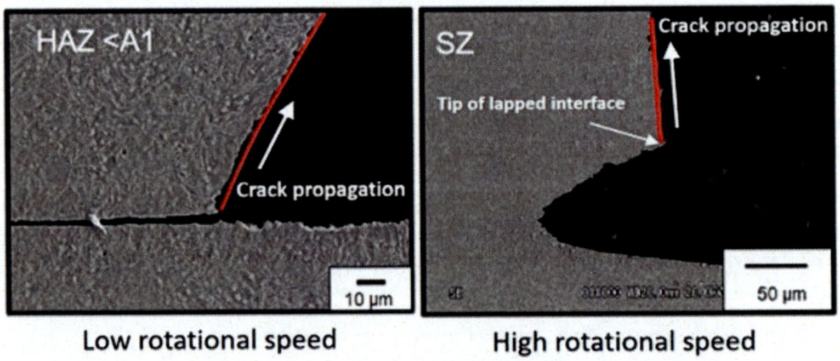

Fig. 6 SEM images of the initiation sites of the welds during tensile shear testing

Summary

FSLW of UHSS was successfully done with the Co-based alloy tool. The lap welds having high fraction of martensite exhibited the low tensile shear load, arising from the brittle fracture. The decrease in rotational speed decreased the brittleness of the stir zone, increasing the tensile shear load. The excessively low rotational speed caused the partly insufficient fragmentation of the lapped interface, reducing the tensile shear load. This study showed that the reduced brittleness of the stir zone with the well fragmentated lapped interface, arising from the low rotational speed, was an effective strategy for the superior mechanical properties of the lap welds of the UHSS.

Acknowledgements The authors are grateful to Mr. A. Furukawa, Mr. K. Sato, and Dr. Z. Lyu for technical assistance. This paper was partially based on results obtained from a future pioneering project commissioned by the New Energy and Industrial Technology Development Organization (NEDO) which is co-research with Innovative Structural Materials Association (ISMA).

References

1. Khan MI, Kuntz ML, Su P, Gerlich A, North T, Zhou Y (2007) Resistance and friction stir spot welding of DP600: a comparative study. Sci Technol Weld Join 12(2):175–182. https://doi.org/10.1179/174329307X159801
2. Hovanski Y, Santella M, Grant GL (2007) Friction stir spot welding of hot-stamped boron steel. Scripta Mater 57(9):873–876. https://doi.org/10.1016/j.scriptamat.2007.06.060
3. Ohashi R, Fujimoto M, Mironov S, Sato YS, Kokawa H (2009) Effect of contamination on microstructure in friction stir spot welded DP590 steel. Sci Technol Weld Join 14(3):221–227. https://doi.org/10.1179/136217108X388642
4. Abdollah-Zadeh A, Saeid T, Sazgari B (2008) Microstructural and mechanical properties of friction stir welded aluminum/copper lap joints. J Alloys Comp 460(1–2):535–538. https://doi.org/10.1016/j.jallcom.2007.06.009
5. Chen YC, Nakata K (2009) Microstructural characterization and mechanical properties in friction stir welding of aluminum and titanium dissimilar alloys. Mater Des 30(3):469–474. https://doi.org/10.1016/j.matdes.2008.06.008
6. Bisadi H, Tavakoli A, Sangsaraki MT, Sangsaraki KT (2013) The influence of rotational and welding speeds on microstructures and mechanical properties of friction stir welded Al5083 and commercially pure copper sheets lap joints. Mater Des 43:80–88. https://doi.org/10.1016/j.matdes.2012.06.029
7. Jana S, Hovanski Y, Grant GJ (2010) Friction stir lap welding of magnesium alloy to steel: A preliminary investigation. Metall Mater Trans A 41A(12):3173–3182. https://doi.org/10.1007/s11661-010-0399-8
8. Ghosh M, Kumar K, Mishra RS (2011) Friction stir lap welded advanced high strength steels: microstructure and mechanical properties. Mater Sci Eng A 528:8111–8119. https://doi.org/10.1016/j.msea.2011.06.087
9. Sato YS, Urata M, Kokawa H (2002) Parameters controlling microstructure and hardness during friction-stir welding of precipitation-hardenable aluminum alloy 6063. Metall Mater Trans A 33A(3):625–635. https://doi.org/10.1007/s11661-002-0124-3

10. Sato YS, Miyake M, Kokawa H, Omori T, Ishida K, Imano S, Park SHC, Hirano S (2011) Development of a cobalt-based alloy FSW tool for high-softening-temperature materials. In: Friction stir welding and processing VI, pp 3–9. Wiley, Hoboken
11. Sutton MA, Yang B, Reynolds AP, Taylor R (2002) Microstructural studies of friction stir welds in 2024–T3 aluminum. Mater Sci Eng A 323(1–2):160–166. https://doi.org/10.1016/S0921-5093(01)01358-2
12. Sato YS, Yamanoi H, Kokawa H, Furuhara T (2007) Microstructural evolution of ultrahigh carbon steel during friction stir welding. Scripta Mater 57(6):557–560. https://doi.org/10.1016/j.scriptamat.2007.04.050
13. ISO 25239-1:2020 Friction stir welding—Aluminium – Part 1: Vocabulary
14. Matsushita M, Kitani Y, Ikeda R, Ono M, Fujii H, Chung Y-D (2011) Development of friction stir welding of high strength steel sheet. Sci Technol Weld Join 16(2):181–187. https://doi.org/10.1179/1362171810Y.0000000026
15. Mironov S, Sato YS, Yoneyama S, Kokawa H, Fujii HT, Hirano S (2018) Microstructure and tensile behavior of friction-stir welded TRIP steel. Mater Sci Eng A 717(21):26–33. https://doi.org/10.1016/j.msea.2018.01.053

Effect of Locally Beta-Transformed Area on Fatigue Crack Propagation Resistance in a FSWed Ti-6Al-4V

M. Okazaki and S. Hirano

Abstract There are growing interest and potential application of FSW in fabricating titanium components in aerospace and space components. In the case of near alpha Ti-6Al-4 V (Ti-6-4) alloy, a locally transformed area from hcp to beta structure is sometimes formed relating to the hot spot in temperature during the FSW process. The purposes of this research are to get basic knowledge on the above questions or dilemma, via exploration on the role of microstructure. Here, special attention has been paid to the effect of the locally beta-transformed microstructures (; LBTM). The experimental study demonstrated that the local fatigue crack propagation (FCP) rates in the stirred zone and near the interfacial zone were higher than those in the base metal by a few times. The local fatigue crack threshold level was also significantly affected. The local beta-transformed microstructure had undesirable effect on the FCP, not only in the longitudinal but in the transverse cracks.

Keywords Joining · Characterization · Titanium

Introduction

Friction stir welding (FSW) technology is relatively new and offers a number of advantages over conventional welding techniques namely higher process speeds, elimination of filler materials, and minimal shrinkage and post weld distortion [1]. The growing interest and potential application of FSW in fabricating titanium components in aerospace and space components entails further work be carried to understand the damage tolerance properties of FSWed joints. Large amount of works has been done to optimise and understand the damage behavior of friction stir welded components [2–8]. The roles of microstructure gradation, hardness distribution, and

M. Okazaki (✉)
Niigata Institute of Technology, Fujihashi 945-1195, Japan
e-mail: m_okazaki@niit.ac.jp

S. Hirano
Formerly Hitachi Research Lab, Hitachi 319-1292, Japan

Y. Hovanski et al. (eds.), *Friction Stir Welding and Processing XII*,
The Minerals, Metals & Materials Series,
https://doi.org/10.1007/978-3-031-22661-8_10

residual stresses have been articulated by a number of researchers working on Titanium alloys FSWs [7–10]. It has been shown that a very complicated and inhomogeneous microstructural developments has been unavoidable to have good joint efficiency. In the case of near alpha Ti-6Al-4 V (Ti-6-4) alloy, a transformation from hcp to beta structure gets significant, depending on the FSW condition; a development of the locally beta-transformed microstructures (LBTM) was also significant inside of the FSWs in some cases. On the other hand, some references have confirmed that fatigue properties of titanium alloy base materials are closely related to their microstructures, and they are not always directly related to static mechanical properties. As an example, Ti-6-4 alloy with a coarse beta structure had higher fatigue crack threshold and higher FCP resistance, than those of the alloy with fine duplex structure, because of surface-induced fatigue crack closure phenomenon which can reduce the FCP driving force [11–13]. Thus, upon optimizing the fatigue design of FSWs of titanium alloys, the questions arise; what is(are) the roles of microstructures on FCP?, and what effect(s) should be taken into account regarding the mixed and transformed microstructure frequently appearing in FSW? The role(s) of LBTM has(have) been still unclear.

The purposes of this research are to get basic knowledge on the above questions or dilemma, via exploration on the role of microstructure and the residual stresses in the local FCP resistance of a Ti-6-4 FSWs, where special attention has been paid to the effect of LBTM and to the crack propagation near the microstructural interfaces.

Experimental Methods

FSW

The starting material in this work was a 7 mm FSWed joint of a Ti-6Al-4 V(Ti-6-4) alloy jointed through a single pass in a butt joint configuration (Fig. 1). The following FSW condition were employed; 120 rpm in tool rotation, 15 mm/min. in welding speed. Here, the welding tool made of a Co-based alloy which shoulder diameter was 12 mm was used [10].

Fatigue Crack Propagation Tests

The mechanical properties; hardness distribution, yield strength, tensile strength, elongation and strain to rupture, were measured by means of miniature specimens with a thickness of 1.0 mm, Fig. 2a, those were extracted by electro-discharge machining (EDM) transverse to the weld direction such that the gage length contained the stirred zone. Taking account of the specifically developed inhomogeneous microstructures in the FSWs, as shown later, these measurements were carried by at

Chemical Composition:

Al	C	Fe	N	O	V	Ti
6.28	0.1	0.21	0.010	0.13	4.05	bal.

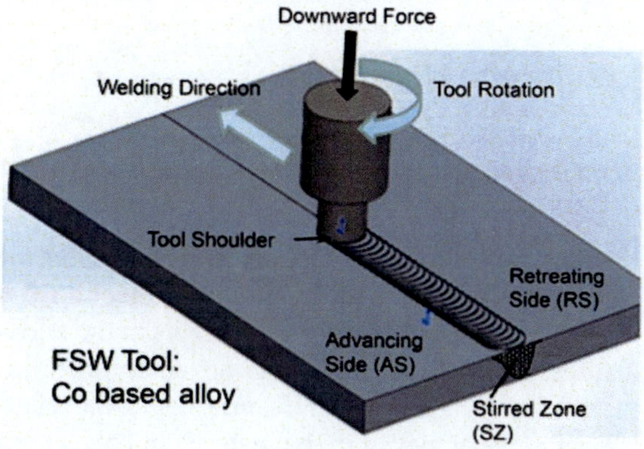

Fig. 1 Basic principle of FSW

the multiple areas, typically at 0.3 and 1 mm site in depth from the top surface of the FSWed plate.

The fatigue crack growth rates in the base metal and welded area were measured by means of compact tension (CT) specimens of 0.8 mm in thickness with initial notches in the base metal and the stirred zone, respectively as shown in Fig. 2b. These CT specimens were precisely extracted at 0.4 mm depth from the top surface of the FSWed block by EDM from both the welded area. A series of CT specimens extracted here have such an orientation that the fatigue cracks propagate to the welding, or longitudinal direction, because of thickness limitation of the Ti-6-4 base plate. The fatigue crack propagation (FCP) tests were carried out under ambient

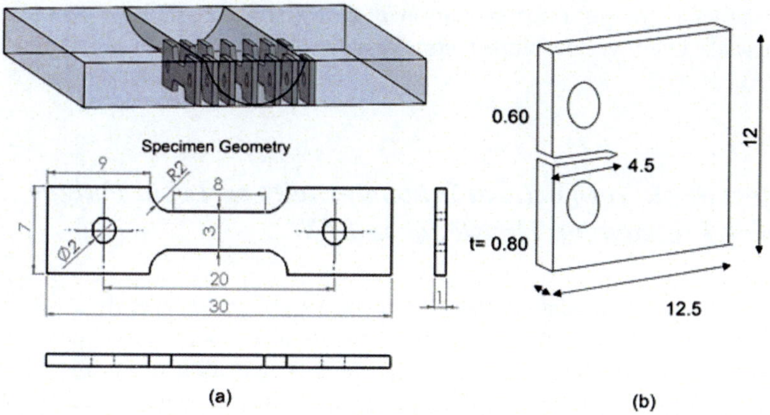

Fig. 2 Method to extract small sample specimens used for **a** tensile and **b** fatigue crack propagation tests

conditions, a load ratio, R, of 0.6 and loading frequency of 20 Hz, by means of a servo-electro-hydraulic machine. The crack length was monitored and measured through a travelling digital microscope continuously focused on the specimen surface.

Results and Discussion

Microstructure, Hardness Distribution and Tensile Strength

Figure 3 presents the microstructure developed in the FSWs, and those were observed on the plane normal to the longitudinal direction of the original 7 mm thickness weld. It is found that refinement of the grains in the stirred zone (SZ) was significant, and that there was a noticeable inhomogeneity in grain size and microstructure through the thickness. The former feature could be related to the high lattice strain and temperatures achieved in the material in contact with the tool shoulder, followed by rapid cooling after the welding. With respect to the latter features, the grain sizes were found to be largest within the SZ at a depth of 0.5 mm from the surface, larger than those at the weld root. It is worthy to note in Fig. 2 that the transformation-induced lamellar structure consisting of alpha/beta bimodal phases, or the formation of the locally beta-transformed microstructures (LBTM) was also significant at the advancing side (AS). This type of microstructure is a direct evidence showing that the local temperature was increased enough during the stirring process so that the transformation of Ti-6-4 alloy was taken place. These characteristic features must be closely related to the low thermal conductivity of titanium alloys and the FSW conditions.

Vickers hardness profiles at different depths across the weld showed that hardness peak appeared in the SZ corresponding to the grain refinement there (Fig. 4). The scatter in hardness was significant on the 3 mm line from the top where the LBTM consisting of the prior non-transformed duplex structure and the beta-transformed coarse structure was developed, corresponding to microstructural inhomogeneity in Fig. 3.

Effect of Microstructural Inhomogeneity on Local Fatigue Crack Propagation Resistance in FSW

By means of the miniature sample specimens extracted from the as-FSWed plate without post weld heat treatment, the local fatigue crack propagation resistance was measured and compared between each crack propagation site in Fig. 5, where the FCP curves are correlated with stress intensity factor range, a traditional fracture mechanics parameter. Regarding as how to extract each specimen, refer to the illustration in Fig. 5 where the AS and RS express the advancing and the retreating side

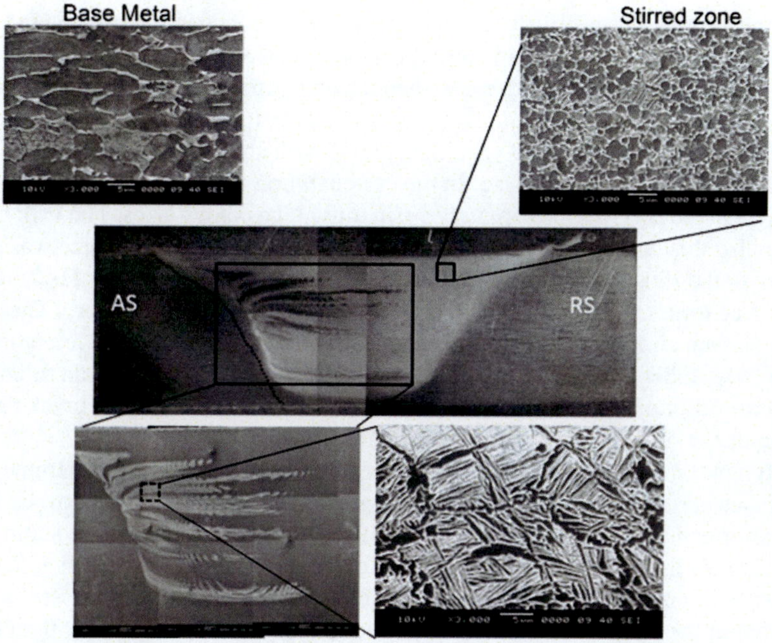

Fig. 3 Microstructure of FSWed Ti-6Al-4 V plate in which a LBTM is nucleated

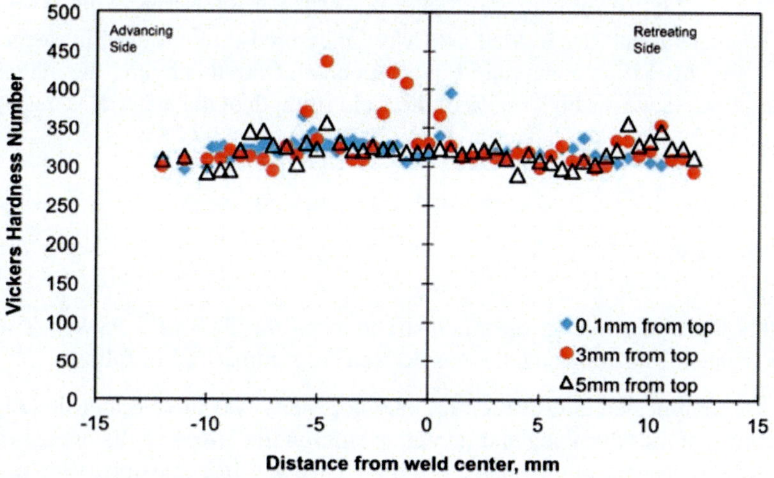

Fig. 4 Hardness distribution of FSWed Ti-6Al-4 V plate

during the FSW, respectively. Comparing with the FCP rates propagating in BM side, all the FCP rates were significantly promoted. It is worthy to note that the rates at the SZ were also accelerated, even when the microstructure refinement was achieved there. The recrystallized equiaxed microstructure in the SZ has been observed to contain significantly refined prior beta grains with alpha colony sizes of one quarter to half the prior beta grain size. In the current work, the prior beta was refined from 13 μm in the BM to approximately 5 μm thus alpha colony sizes. The FCP behavior was also the case in the interfacial zone (IZ) site where the grain size was smaller; refer to the illustration of Fig. 5. One of exception was seen at the HAZ site those was free from mechanical stirring but was subjected to thermal history. These trends can be seen almost in common independently on the advancing and retreating sides, and independently on the depth from the FSW surface. It is important to note again that the difference in FCP resistance was noticeable near the fatigue crack threshold, or under lower stress intensity factor range.

It must be meaningful to investigate the effect of the LBTM on the fatigue fracture resistance, because it is not hard to distinguish the fracture surface between the non-transformed and the transformed area. This is the case not only in the static fracture surface but in the fatigue crack propagation test (Fig. 6). Looking at Fig. 5, it seems that the fatigue crack propagation resistance at the "IZ on AS" was almost comparable to that at the "IZ on RS", and both of them were significantly lower than the base metal. Hence, the stirring process itself would play a role in reducing the fatigue crack propagation resistance under the present FSW condition.

The residual stress may also affect the FCP behavior. The residual stresses in the base metal and the welded area were measured through the $\mathrm{Sin}^2\psi$ method of the X-ray diffraction technique: Fig. 7. As noted from this figure, the magnitude of residual stress was as high as 5% of the yield strength of the material. Thus, it played a secondary role in the reduction of the FCP resistance by FSW.

Summary

In this work, the role of the microstructure on the FCP of a FSWed Ti-6-4 was investigated. The main conclusions derived are summarized as follows:

(1) A complicated microstructural inhomogeneity was three-dimensionally developed at the advancing and retreating sides in the Ti-6-4 FSWs, which consisted of the primary alpha phase, and the coarse and fine transformation associated with the locally beta-transformed microstructures (LBTM).

(2) The local FCP rates in the stirred zone and the interfacial zone were higher than those in the base metal and heat affected zone by a few times. The local fatigue crack threshold level was also significantly affected. Here, the role of microstructure was more significant than that of residual stress.

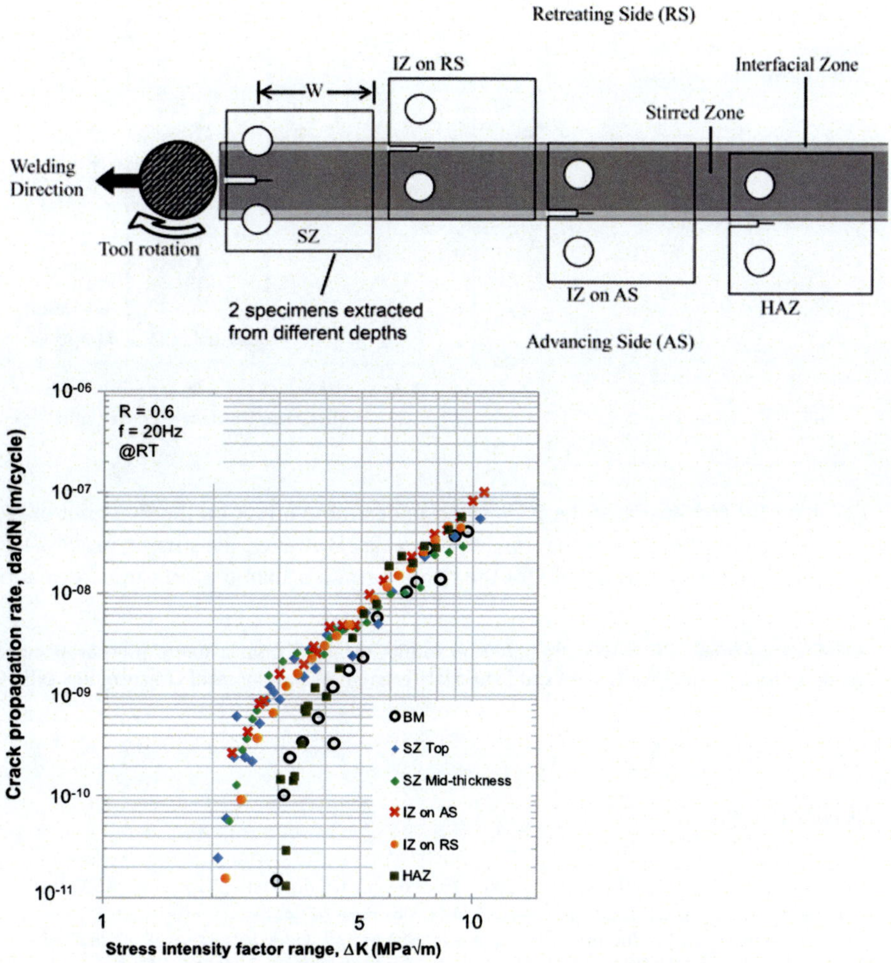

Fig. 5 Fatigue crack propagation resistance depending on the local microstructures in the FSWed plate

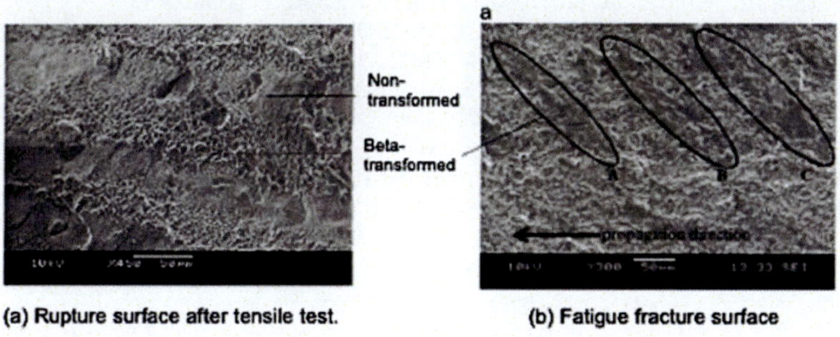

Fig. 6 Tortuous fracture path around the LBTM area

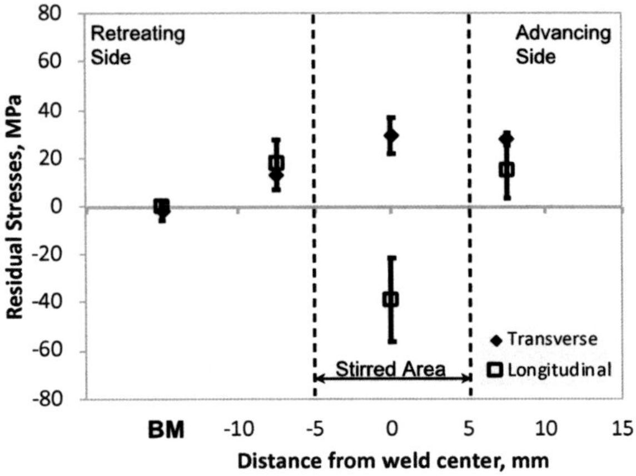

Fig. 7 Residual stress distribution measured by X-ray diffraction method

(3) The LBTM inside of the FSWs had undesirable effect on the FCP, not only in the longitudinal but in the transverse cracks. However, the stirring process itself must be more responsible for the behavior rather than the beta-transformation.

Acknowledgements This work is based on results obtained from a future pioneering project commissioned by the New Energy and Industrial Technology Development Organization (NEDO).

References

1. Mishra RS, Mahoney MW (eds) (2007) Friction stir welding and processing. ASM Inter, 1–5
2. Nandan R, DebRoy T, Bhadeshia HKDH (2008) Prog Mater Sci 53:980–1023
3. Sato Y, Kokawa H, Enomoto M, Jogan S (1999) Metall and Mater Trans A 30:2429–2437
4. Brown R, Tang W, Reynolds AP (2009) Mat Sci Eng: A, 513–514:115–121
5. John R, Jata KV, Sadananda K (2003) Int J Fatigue 25:939–948
6. Pasta S, Reynolds P (2005) Fatigue Fracture Eng Mat Struct 31:569–580
7. Edwards PD, Ramulu M (2009) Sci Tech Weld Join 14(5):476–483
8. Pilchak AL, Williams JC (2011) Metallurgical Trans 42A:1630–1645
9. Pao PS, Fonda RW, Jones HN, Feng CR, Moon DW (2009) In: Mishra RS, Mahoney MW (eds) Friction stir welding and processing V. TMS, (2009), pp 29–38
10. Muzvidziwa M, Okazaki M, Suzuki K, Hirano S (2016) Mater Sci Eng A652:59–68
11. Lutjering G (1998) Mat Sci Eng A243:32–45
12. Suresh S (1983) Met Trans 14A:2375–2385
13. Minakawa K, McEvily AJ (1981) Scripta Met 15:633–636
14. Murukami Y et al (1986) Stress intensity factor handbook. The Soc Mater Sci Japan

Part V
Spot Technologies

Joining of High Strength Low Ductility AA7055 by Friction Self-piercing Rivet

Yong Chae Lim, Hui Huang, Yiyu Wang, Yuan Li, and Zhili Feng

Abstract High strength aluminum (Al) alloy is one of higher specific strength materials for decarbonization in transportation industries. Because of low ductility at room temperature, conventional mechanical fastening such as self-piercing riveting produces cracks at the joint. In this work, we applied friction self-piercing riveting to join Al alloy (AA) 7055. No cracks were observed in the joints because of the improved local ductility of Al alloy by the generated frictional heat during joining step. Numerical modeling of joining process was applied to guide rivet geometry design and rivet material strength. Mechanical integrity of the AA7055 joints was assessed by lap shear tensile and cross-tension testing. Metallurgical characterizations revealed solid-state bonding formed not only between the rivet and surround Al materials, but also upper and lower Al sheets at the joint interface. Both solid-state bonding and mechanical interlocking between the flared rivet and bottom AA7055 sheet were the major joint mechanisms.

Keywords Friction self-piercing riveting · High strength 7xxx Aluminum alloys · Mechanical joint properties · Solid-state joining

---

Y. C. Lim (✉) · H. Huang · Y. Wang · Y. Li · Z. Feng
Materials Science & Technology Division, Oak Ridge National Laboratory, One Bethel Valley Road, Oak Ridge, TN 37831, USA
e-mail: limy@ornl.gov

H. Huang
e-mail: huangh@ornl.gov

Y. Wang
e-mail: wangy4@ornl.gov

Y. Li
e-mail: liy5@ornl.gov

Z. Feng
e-mail: fengz@ornl.gov

Y. Hovanski et al. (eds.), *Friction Stir Welding and Processing XII*,
The Minerals, Metals & Materials Series,
https://doi.org/10.1007/978-3-031-22661-8_11

Introduction

High strength aluminum alloy (AA) is considered as a one of higher specific strength materials for lightweight vehicle applications. 7xxx AA (Al-Zn-Mg-Cu) has much higher strength than the 5xxx and 6xxx Al alloys, leading to great potential for decarbonization in vehicle applications. One of technical obstacle to prevent widespread of 7xxx AA for transportations is lack of suitable joining process. For example, fusion welding of 7xxx AA shows a solidification cracking because of different solidification rates between copper-rich zone and copper-free zone [1]. Another example is a cold cracking problem when mechanical fastening (e.g., self-piercing riveting (SPR)) is used at room temperature [2]. This is mainly due to low ductility at room temperature. Therefore, additional auxiliary heating system and joining step are required, increasing joining process time and cost.

To alleviate the above issues, solid-state-based joining processes, such as friction stir welding (FSW) or refill friction stir spot welding were used for joining of 7xxx Al alloy by avoiding melting of material with relatively low heat input compared with conventional fusion welding [3, 4]. However, strength loss in the thermomechanical affected zone and heat affected zone (HAZ) is a still problem. Recently, friction-based self-piercing riveting process is developed for joining of low ductility lightweight materials such as magnesium alloy, 7xxx AA, and carbon fiber reinforced composites without cracking issue [5–10].

In this work, we applied F-SPR process to join AA7055-T76 to AA7055-T76 by varying process conditions such as spindle rotational speeds and axial plunge depths. Mechanical joint strength of F-SPR joints was evaluated by lap shear tensile testing. Cross-sectional analysis for F-SPR joint was conducted to study joint formation. Vickers microhardness measurement was used to assess the local hardness evolution due to frictional heat and shear stress during F-SPR process.

Materials and Experimental Methods

Materials

In this work, 2.5 mm thickness of high strength AA 7055-T76 sheets was used to spot join by F-SPR process. Chemical compositions and mechanical properties of the AA7055-T76 are summarized in Tables 1 and 2. For a consumable rivet, 1018 low carbon steel was employed to fabricate the hexagonally designed rivet head (9.535 mm width) to be externally driven by the tool holder during joining process. The rivet diameter (6.8 mm) and leg length (6 mm) were customized for the selected Al alloy. For back supporting anvil, a flat die with 1.5 mm cavity depth was used to induce material flow and produce mechanical interlocking between the rivet and bottom Al sheet.

Table 1 Summary of chemical composition of AA7075-T76 [11]

Element (wt.%)	Zn	Mg	Cu	Zr	Cr	Si	Fe	Mn	Ti	OE	OT	Remainder
AA7055-T76	7.6–8.4	1.8–2.3	2.0–2.6	0.08–0.25	0.04 max	0.10 max	0.15 max	0.05 max	0.06 max	0.05 max	0.15 max	Al

Table 2 Material properties of AA7055-T76 [1]

Material properties	Yield strength (MPa)	Tensile strength (MPa)	Elongation (%)
AA7055-T76	586	613.6	7

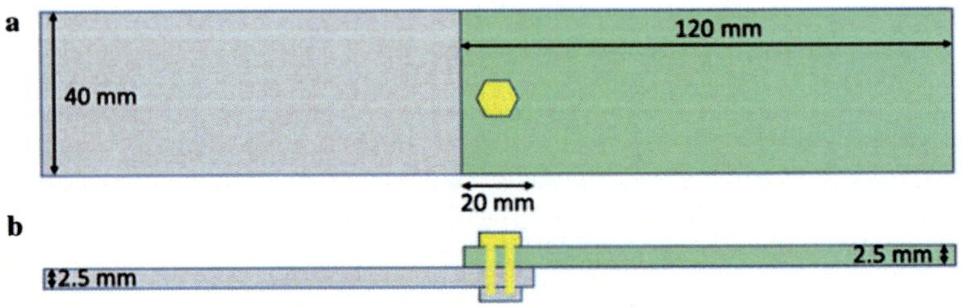

Fig. 1 Schematic of lap shear tensile coupon. **a** top view, **b** side view

AA7055 sheets were cut into lap shear coupons (40 mm wide and 120 mm long), as depicted in Fig. 1. Al surfaces were cleaned by acetone to remove any greases and dusts before joining.

Description of F-SPR Process

A description of the F-SPR process can be found in the authors' prior publications [9, 10]. A brief explanation of the joining process is provided for readers. As illustrated in Fig. 2, a spinning rivet with an axial downward force is plunged into stacked Al sheets. During this step, local frictional heat is produced between the rivet and the bottom Al sheet metal, resulting in improved local ductility of the AA7055-T76. Finally, the rivet leg flares out in the bottom Al sheet based on the supporting die geometry, leading to mechanical interlocking in the workpiece. F-SPR process was initially developed by varying spindle rotation speeds and axial plunge depths at fixed axial plunge speed of 2.86 mm s^{-1} that is a maximum speed of the current our

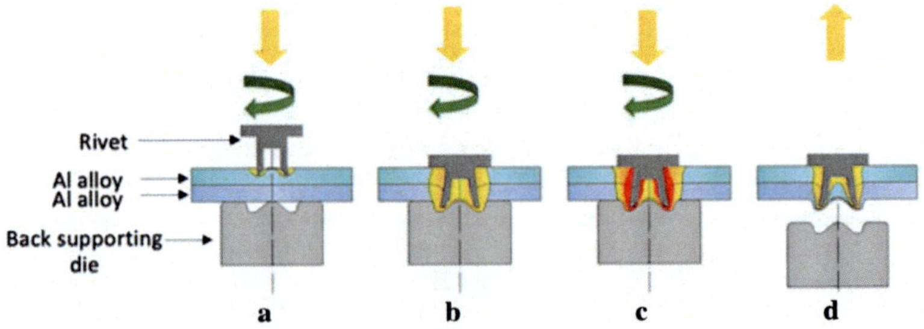

Fig. 2 Schematic of F-SPR process

joining machine. The F-SPR process conditions were optimized by achieving the highest lap shear fracture load and no cracking on the backside of the AA7055-T76.

Description of Lap Shear Tensile Testing

Lap shear tensile testing was conducted to assess the joint strength of F-SPR specimens by an MTS Systems tensile machine with a constant crosshead speed of 10 mm min^{-1} at room temperature. Spacers were used to hold the lap shear coupons to align them vertically between the grips during lap shear tensile testing.

Metallography Characterizations

The joint formation between the rivet and Al sheet was characterized by the cross-sectioned F-SPR coupon. The sample was cross-sectioned, mounted, grinded, and polished for metallographic characterization using standard metallographic processes. Then, an optical microscope was used to measure characteristic length of the joint and any metallurgical bonding during joining process. A microhardness tester (Leco LM 100AT) was used for Vickers microhardness measurement. Each welded sample was tested under following conditions: 200 μm spacing, 100 g of load, and 10 s of dwell time.

Results and Discussion

Lap Shear Tensile Strength of F-SPR Joints

Load versus displacement curves from lap shear tensile testing are plotted in Fig. 3. The average peak failure load was found to be 11.70 ± 0.17 kN. Displacement at failure is ranged from 6 mm to 6.6 mm, indicating more ductile failure than the brittle mode. Figure 4 shows representative fractographic image from lap shear tensile testing of F-SPR joint. The rivet pulled out the bottom sheet of AA7055-T76, as shown in Fig. 4a. Cavity in the rivet was filled with upper and lower sheet of Al alloy. Also, some Al materials retained at the periphery of the rivet after lap shear tensile testing, resulting in partial Al sheet fracture at the periphery of the joint. The strong mechanical joint strength can be attributed not only to the mechanical interlocking of the flared rivet with the bottom aluminum sheet, but also to the solid-state joining of the rivet with the surrounding aluminum sheet. Further metallurgical characterizations were performed to study the joint mechanism in the next section.

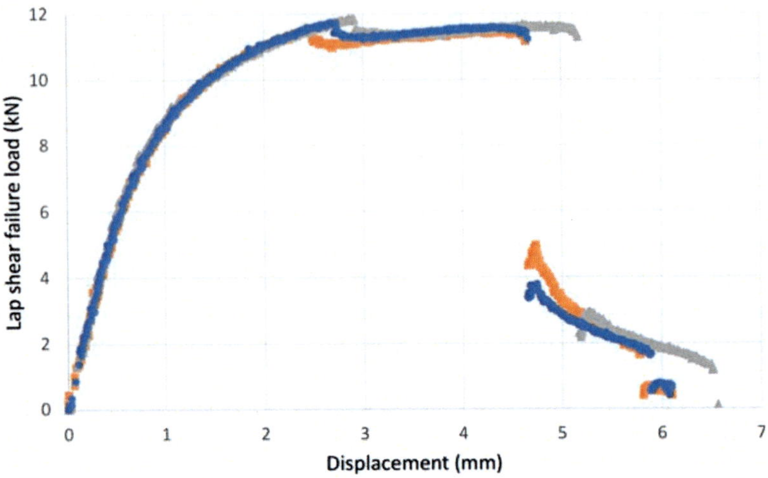

Fig. 3 Load and displacement curves for lap shear tensile tested F-SPR AA7055-T76 joints

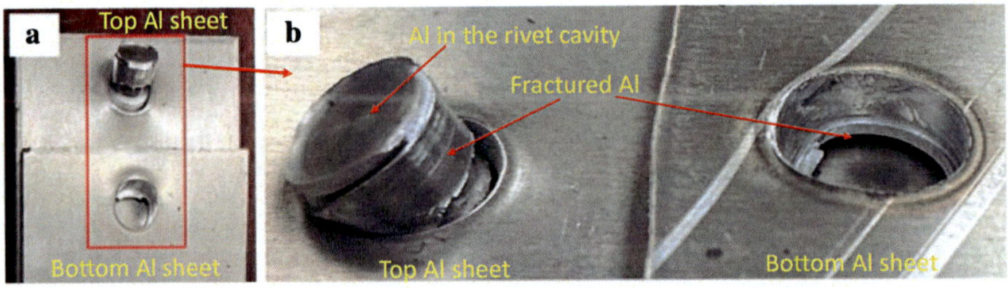

Fig. 4 (**a**) Fractured F-SPR joints after lab shear tensile testing (**b**) magnified fractography, showing Al pullout failure mode

Metallurgical Characterizations

Figure 5 provides a cross-sectional view of the F-SPR Al joint. Mechanical interlocking distance (two red lines on each side) was measured as 0.4 mm. Mechanical interlocking size determines the level of locking strength between the rivet and the bottom sheet, along with the load path within the joint. Friction heat generated during F-SPR improves local ductility of Al alloy, so no cracking of AA7055 was observed in Fig. 5(a). It is found out that ductility of AA7075-T6 is increased at temperature ranged from 150 to 250 °C [12]. In addition, Ying et al. applied thermal assisted SPR for AA7075-T6 [13]. Maximum mechanical joint strength was achieved without crack on the Al alloy when temperature is ranged from 400 to 450 °C. Figure 5b, c shows a solid-state bonding between upper Al and lower Al sheets at both left (red dot circle) and right sides caused by the frictional and downward axial force. Measured

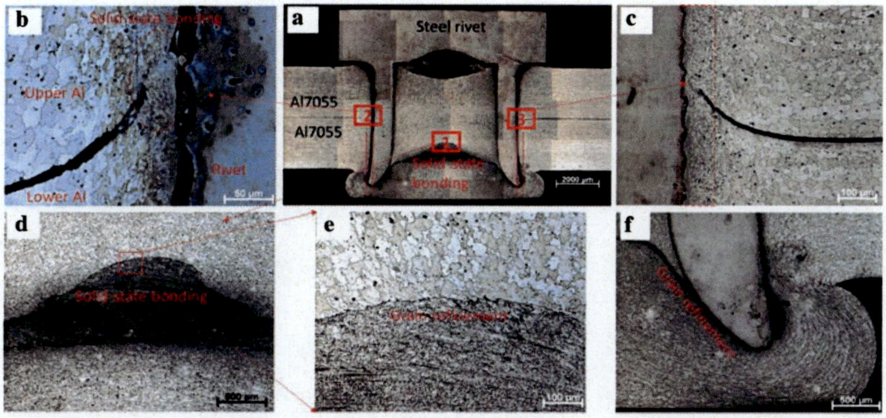

Fig. 5 Optical image of cross-sectioned F-SPR AA7055 joint at different joint locations. **a** Low magnification optical image, showing mechanical interlocking between the rivet and bottom Al sheet. No cracking of bottom Al sheet was observed. **b** Magnified optical image at left side joint interface, showing solid-state bonding between upper and lower Al sheets (red dot circle). **c** Magnified optical image at the right side joint interface, showing grain refinement of Al sheets near joint interface (red dot box). **d** Solid-state joining between upper and lower Al sheets at the rivet cavity. **e** Magnified optical image showing grain refinement. **f** Grain refinement of Al and material flow at the right side rivet tip

solid-state joining size is approximated 30 µm. Also, F-SPR process induces Al material flow near the joint interface, forming a hook shape. Next, grain refinement of Al near the joint is seen due to dynamic recrystallization. Figure 5d presents a solid-state bonding between upper and lower Al sheet at the rivet cavity. This solid-state joining is facilitated by spinning motion of rivet under frictional heat and compression load during F-SPR process. Figure 5e shows grain refinement of upper Al sheets where solid-state joining has taken place with lower Al sheets. Finally, material flow and grain refinement at the rivet tip is presented in Fig. 5f. Again, plastic deformation of lower Al sheet under large shear motion and frictional heat during F-SPR induces such material flow and refinement of grain in the lower Al sheet.

Vickers microhardness mapping of steel rivet and AA7055-T76 in the F-SPR joint is presented in Fig. 6a. The line scans (two dot blacklines) of the steel rivet and Al alloy where $Y = 4$ mm and 2.6 mm are plotted in Fig. 6b. The average Vickers microhardness of steel rivet is approximately 280 HV. There is no significant reduction of hardness for the steel shank because frictional heat did not cause high temperature to soften the steel rivet. From open literature, it is found that mechanical properties (i.e., yield and tensile strength) for low carbon steel (ASTM 1022) starts reducing above 400 °C [14]. Therefore, it is rational that estimated peak temperature during F-SPR is less than 400 °C.

The Vickers microhardness of as-received AA7055-T76 is measured to be ranged from 180 to 195 HV. The microhardness of AA7055-T76 near the steel rivet-Al alloy joint interface is about 230 HV which is much higher than the base material. This

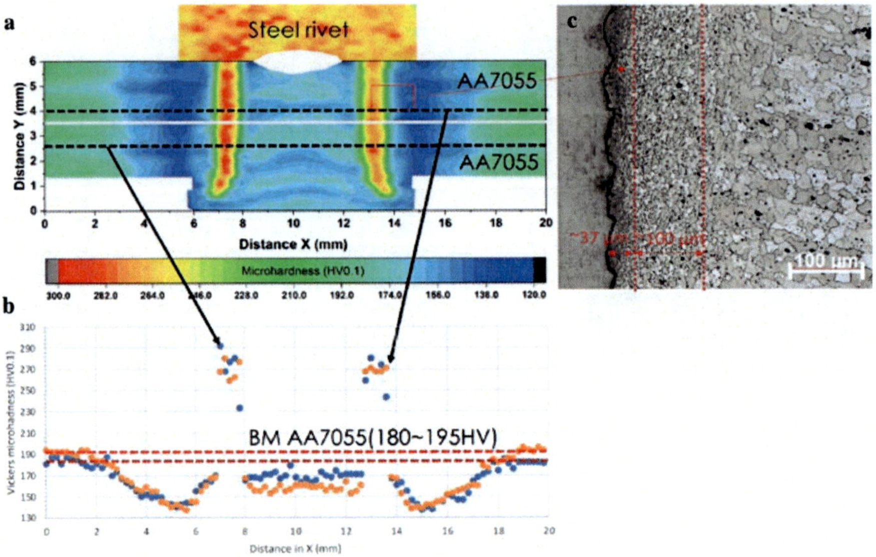

Fig. 6 (**a**) Mapped Vickers microhardness measurement for F-SPR Al joint (**b**) line plotted Vickers microhardness at Y distance = 4 mm and 2.6 mm in **a**. (**c**) optical micrograph at the right side joint, showing grain size gradient. Refined grain near the joint interface was found, and then grain size increased as moved toward to right direction

higher hardness results from the grain refinement caused by dynamic recrystallization during F-SPR process, as shown in Fig. 6c [red dot box]. Refined grain zone is approximately 130 μm. Above this distance, there is a local reduction of microhardness caused by frictional heat during F-SPR. This soft zone is about 2 mm long. Then, microhardness is back to the 180–195 HV again. This local hardness drop zone (i.e., HAZ) can be corelated to the fracture location of F-SPR joints from lap shear tensile testing. Estimated peak temperature below 400 °C is not only high enough to increase local ductility to avoid cold crack, but also to induce local hardness drop at the joint location. For this reason, weakest location is at the HAZ based on the current F-SPR joint process parameters.

Conclusions

F-SPR process was developed to spot join high strength low ductility AA7055-T76. Frictional heat generated during F-SPR process improved local ductility of Al alloy, resulting in crack free joint. The averaged lap shear fracture load is 11.7 kN with failure of the bottom Al sheet, indicating strong mechanical interlocking between the flared rivet and bottom Al sheet. From the cross-sectional characterization, solid-state joining between upper Al and lower Al sheet at the joint interface was found.

Due to frictional heat and high shear motion, grain refinement of Al alloy near the joint interface was also observed. Vickers microhardness measurement reveals the local hardness drop of Al alloy near the joint due to frictional heat during F-SPR.

Acknowledgements The authors acknowledge the financial support of the US Department of Energy Office of Energy Efficiency and Renewable Energy, Vehicle Technology Office, as part of Joining Core Program. Oak Ridge National Laboratory is managed by UT-Battelle, LLC for the US Department of Energy under Contract DE-AC05-00OR22725. The authors thank Russell Long at Arconic for providing materials.

References

1. Temmar M, Hadji M, Sahraoui T (2011) Effect of post-weld aging treatment on mechanical properties of tungsten inert gas welded low thickness 7075 aluminium alloy. Joints Mater Des 32(6):3532–3536
2. Jäckel M, Grimm T, Landgrebe D (2016) Approaches for mechanical joining of 7xxx series aluminum alloys. AIP Conf Proc 1769:100010
3. Rajakumar S, Muralidharan C, Balasubramanian V (2011) Influence of friction stir welding process and tool parameters on strength properties of AA7075-T6 aluminium alloy joints. Mater Des 32(2):535–549
4. Shen Z, Yang X, Zhang Z, Cui L, Li T (2013) Microstructure and failure mechanisms of refill friction stir spot welded 7075–T6 aluminum alloy joints. Mater Des 44:476–486
5. Li Y, Wei Z, Wang Z, Li Y (2013) Friction self-piercing riveting of aluminum alloy AA6061-T6 to magnesium alloy AZ31B. J Manuf Sci Eng 135(6):061007
6. Liu X, Lim YC, Li Y, Tang W, Ma Y, Feng Z, Ni J (2016) Effects of process parameters on friction self-piercing riveting of dissimilar materials. J Mater Proc Technol 237:19–30
7. Lim YC, Warren CD, Chen J, Feng Z (2019) Joining of lightweight dissimilar materials by friction self-piercing riveting
8. Ma YW, Li YB, Lin ZQ (2019) Joint formation and mechanical performance of friction self-piercing riveted aluminum alloy AA7075-T6 joints. J Manuf Sci Eng 141:041005
9. Lim YC, Chen J, Jun J, Leonard D, Brady MP, Warren CD, Feng Z (2021) Mechanical and corrosion assessment of friction self-piercing rivet joint of carbon fiber-reinforced polymer and magnesium alloy AZ31B. J Manuf Sci Eng 143(3):031006
10. Lim YC, Jun J, Leonard D, Li Y, Chen J, Brady MP, Feng Z (2022) Study of galvanic corrosion and mechanical joint properties of AZ31B and carbon-fiber-reinforced polymer joined by friction self-piercing riveting. J Magnes Alloy 10(2):400–410
11. Alcoa aerospace technical fact sheet "Alloy 7055-T74511 and 7055-T76511 extrusions
12. Huo W, Hou L, Zhang Y, Zhang J (2016) Warm formability and post-forming microstructure/property of high-strength AA 7075–T6 alloy. Mater Sci Eng A 675:44–54

13. Ying L, Gao T, Dai M, Hi P, Dai J (2021) Towards joinability of thermal self-piercing riveting for AA7075-T6 aluminum alloy sheets under quasi-static loading conditions. Int J Mech Sci 189:105978
14. Nikulin SA, Rogachev SO, Nikolaev YA, Vasiliev SG, Belov VA, Turilina VY (2021) High-temperature mechanical properties of low-carbon steel used for the manufacture of core catcher vessel. Prog Nucl Energy 142:104015

2D Axisymmetric Modeling of RFSSW Repair and Experimental Validation

Evan Berger, Michael Miles, Paul Blackhurst, Ruth Belnap, and Yuri Hovanski

Abstract Refill friction stir spot welding (RFSSW) is a solid-state spot joining technique that can be used to weld aluminum alloys in various thickness and alloy combinations. From a production perspective, there are times when RFSSW can be terminated prior to completion, leaving an area in need of repair. It is critical to identify the unique conditions that yield excellent bonded properties and the ability to repair a joint that was interrupted during the welding process. Qualified repair techniques are critical for successful implementation of a welding process for use on large weldments with a significant number of spot joints. The focus of this work is to demonstrate a repair technique for RFSSW that can be validated both experimentally and numerically. Modeling is performed using a 2D axisymmetric, thermo-mechanically coupled model, which has previously been validated for the RFSSW joining process. Repaired properties are shown to exceed 90% of the original mechanical properties of the RFSSW process.

Keywords Refill friction stir spot welding · Model · Repair · Thermal comparison · Intermetallic bonding

Introduction

Refill friction stir spot welding (RFSSW) is a variant of friction stir spot welding (FSSW) that was developed to create a flush surface on the top of the weld and to improve strength compared to FSSW [1–3]. Initial numerical simulation of RFSSW started as early as 2006 [4], evaluating the initial deformation induced from the pin coming into contact with the workpiece.

E. Berger · M. Miles · P. Blackhurst · R. Belnap · Y. Hovanski (✉)
Department of Manufacturing Engineering, Brigham Young University, Provo, UT 84602, USA
e-mail: yuri.hovanski@byu.edu

M. Miles
e-mail: mmiles@byu.edu

Y. Hovanski et al. (eds.), *Friction Stir Welding and Processing XII*, The Minerals, Metals & Materials Series,
https://doi.org/10.1007/978-3-031-22661-8_12

From a production perspective, FSSW welds can sometimes be terminated prior to completion, when certain machine thresholds are reached, such as the force threshold. If the machine measures a force above the maximum force threshold, the machine will abort the weld during the welding process and rip out the material under the shoulder and pin, leaving an interrupted weld with a hole.

Interrupted welds can also occur because of the phenomenon of intermetallic bonding between the steel tooling in the FSSW process and the aluminum workpiece. At elevated temperatures of 400 °C and above, the iron in the steel and the aluminum bond to produce an intermetallic compound layer of $FeAl_3$ which covers the steel surface [5, 6]. This phenomenon is exclusive to steel tooling and does not occur with RFSSW processes using tungsten carbide tooling.

Various methods of friction stir welding (FSW) have been used to repair welds as shown in prior work [7, 8], such as repetitive FSW, manual fusion welding, fusion filling plus FSW, and solid-state filling plus FSW. Many methods used a scrolled groove tool without a probe that could produce a material flow towards the rotating center [9].

Self-refilling friction stir welding has been researched as an effort to repair keyhole, where microstructural observation showed that the grains in the refilled zone were significantly refined by the tool [10]. There is still no research on the viability of using refilling techniques as a valid method of repairing interrupted welds.

The current work shows that interrupted welds can be repaired by the RFSSW process using an aluminum plug of similar material in the void. The process refills the void completely, resulting in a smooth surface finish and improved mechanical properties from typical RFSSW processes. This method of RFSSW repair is validated both experimentally and numerically.

Materials and Methods

RFSSW experiments were performed with similar material sheets of 1.50 mm AA2029-T8, a material widely used in supersonic aircraft skin and structural members. The chemical composition of AA2029 is 6% copper, 0.3% iron, and 0.02% magnesium. These composition values are nominal. For the T8 temper, the material was solution heat treated, cold worked, then artificially aged [11].

The tooling setup for the RFSSW repair processes includes a pin, a shoulder, and a rigid clamp. The material sheets rest on the anvil of the machine. The clamp, shoulder, and pin are all concentric and allow for kinematic motion of the rotating pin and shoulder with the non-rotating clamps such as the movement depicted in Fig. 1. All experimentation included in this study included tool speed of pin and shoulder of 2800 rpm in the same direction. These parameters were developed and documented previously as a means of significantly reducing the process cycle time of RFSSW [12, 13]. The five-step process highlighted in Fig. 1 starts with the non-rotating clamp descending until the rotating pin and shoulder both contact the upper surface of the plug. The second step is the compressing of the plug, where both the shoulder and the pin rotate while lowering the plug into the interrupted weld until the tooling is aligned with the rigid clamp. The third step allows the rotating shoulder to plunge into the workpiece, while the rotating pin rises into the tool set, creating a space for the material being displaced by the shoulder. The fourth step returns the shoulder and the pin to the surface of the material, where the pin displaces the material back under the shoulder. The final step shows the tooling retracting completely from the workpiece (Fig. 2).

The geometries of the tools used for the experiments are replicated exactly in the model, as any small deviations in these designs would result in changes in material deformation, heat generation, and forces experienced by the tooling in the model prediction. The tools are made of H13 steel, and the key dimensions are shown in Table 1.

The equipment used for the prediction of RFSSW and data acquisition throughout the experiments was a Bond Technologies RFSSW end-effector, Fig. 3. The machine outputs the positions of the tools and the forces on the tools at any given time.

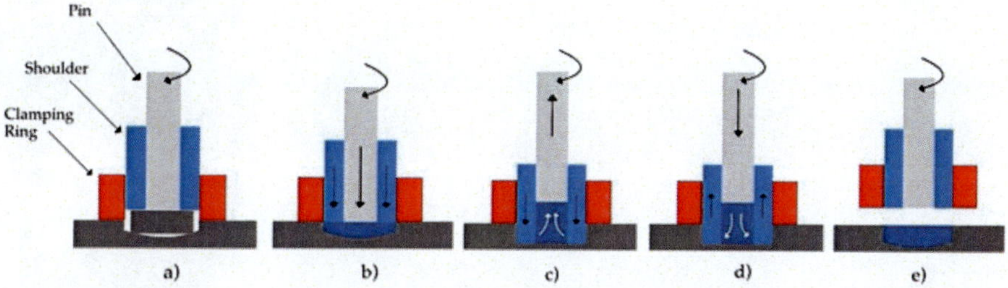

Fig. 1 Schematic of the RFSSW repair process: **a** initial state, **b** clamping phase, **c** plunging phase, **d** refill phase, **e** retraction phase

Fig. 2 RFSSW tooling, including a clamp (left), shoulder (center), and pin (right)

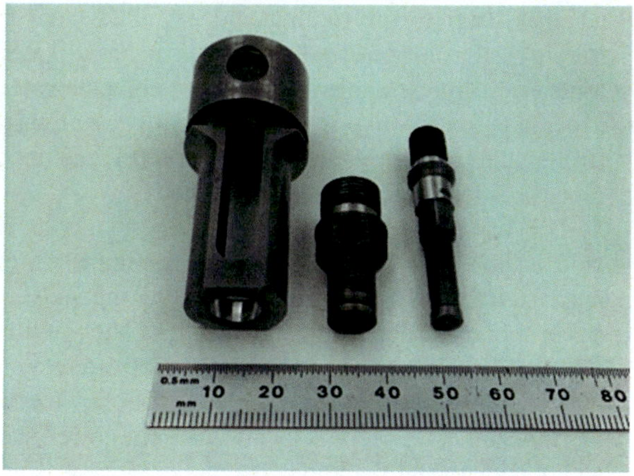

Table 1 Tool dimensions used for numerical model and validation

Tool	Dimension (mm)
Pin outer diameter	5.90
Shoulder inner diameter	6.00
Shoulder outer diameter	9.00
Rigid clamp inner diameter	9.10
Rigid clamp outer diameter	15.00

Fig. 3 Bond technologies RFSSW end-effector

The ferroaluminum compound $FeAl_3$ created during some interrupted welds, attaches itself to the tooling and is pulled from the material workpiece. From a tribological point of view, a layer of material is deformed severely by the tool and transferred into a nanosized grain state at elevated temperature. This creates the necessary conditions for shear instability of a plasticized metal layer. The result is strong adhesion of this metallic layer to the pin and shoulder tooling of the RFSSW process as shown in Fig. 4.

This leaves a concave void, resulting in a volumetric loss in the workpiece. This void is filled with small plugs of aluminum which are then mixed into the base material using RFSSW, thus eliminating the void.

To measure the temperatures during the welding process, five K-type thermocouples are embedded into the lower of the two sheets of AA 2029. This is done by CNC-milling paths for the thermocouple wires as shown in Fig. 5, attaching the thermocouples in precise locations in the lower sheet to gather thermal data inside and outside of the stir zone. The upper sheet was unaltered.

Fig. 4 Aborted weld, with ferroaluminum compound attached to the tooling

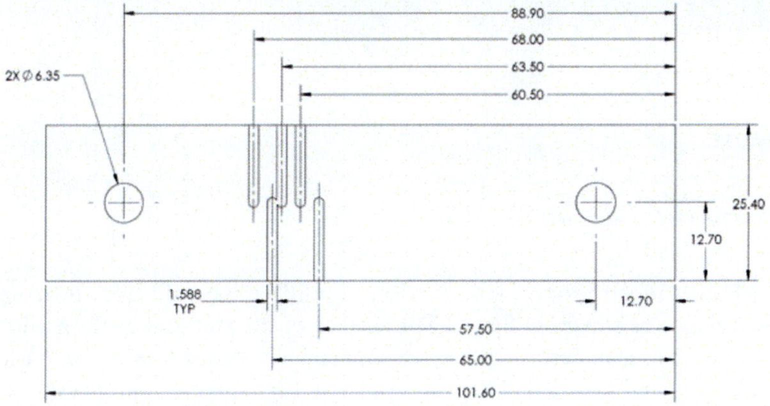

Fig. 5 Thermocouple layout for CNC mill of AA 2029 coupon

The finite element method (FEM) modeling software used to simulate the RFSSW repair process is ForgeNxt 3.2 developed by Transvalor SA. Because the RFSSW tooling has rotational symmetry, a 2D axisymmetric approach was used for modeling, with a Lagrangian description of material flow to manage the non-steady state nature of the process. The computational particulars of the simulation and modeling software are detailed in a previously published article [14]. The material file used for the AA2029 and the method for measuring flow stress are thoroughly outlined in a previously published thesis [15].

The thermocouples are labeled 1–5, from the center of the weld to the outside of the stir zone for both the welding process and the simulation as shown in Fig. 6.

The tools in the model are piloted using rigid contours, which provide velocity boundary conditions according to the kinematics programmed into the simulation. The contours also assume an adiabatic thermal boundary condition because

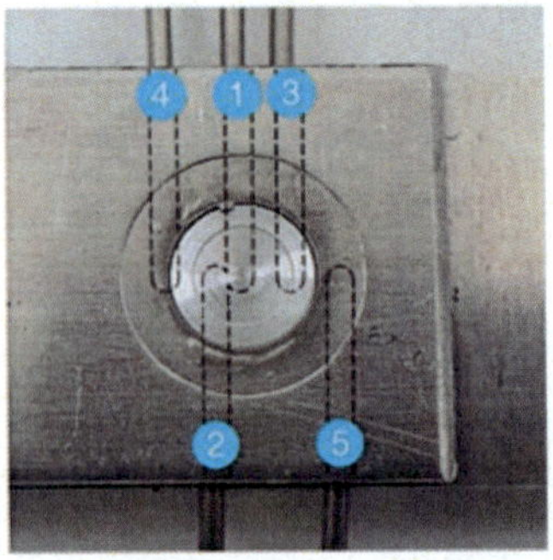

(a)

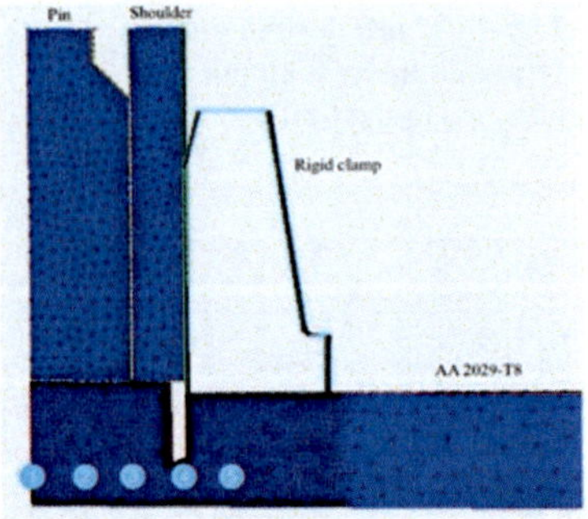

(b)

Fig. 6 Thermocouple layout for the **a** welding process and the **b** model

the welding cycle is so short, and the heat transfer between sheets and clamps is negligible.

Results

Experimental Results

The RFSSW repair process is successful in refilling the void from interrupted and aborted welds as shown in Fig. 7. The surface of the stir zone is flush with the top surface of the base material, with only a slight material rise between the outside of the shoulder tool and the rigid clamp. As shown in Fig. 7, there is proper consolidation during the joining process.

Fig. 7 Cross section of the RFSSW repair weld

Table 2 Tensile strength comparison

	Control	RFSSW repair
1	10,222 N	10,253 N
2	10,680 N	10,587 N
3	10,689 N	11,605 N
4	7469 N	10,952 N
5	10,573 N	9128 N
Mean	9927 N	10,505 N
Standard deviation	1387 N	919 N

Tensile strength testing took place with five RFSSW repair welds and five control RFSSW welds. The results in Table 2 show that the average RFSSW repair is 578 N stronger in tension. These results show that the RFSSW repair process not only completes the welding process but increases the mechanical properties of the weld. These experimental results validate the process as a successful repairing process, both mechanically and aesthetically.

Numerical Results

The temperature comparison between the experimental validation and the simulation differs by 8% at most for the entire duration of the weld. The accuracy of the predictions indicates that the heat generation and the material flow in the model resemble the actual welding repair process. Temperature peaks for the five thermocouples and numerical sensors in the model are shown in Fig. 8.

These temperature predictions give confidence that the material flow generated by the model is reasonably accurate. The model considers the heat generation from both friction and material deformation during the RFSSW repair process. Figure 9 shows the temperature evolution occurring in the material as heat is generated by the high rotational speed of the tools. The heat generation in the model is consistent

Fig. 8 Temperature peaks for each thermocouple

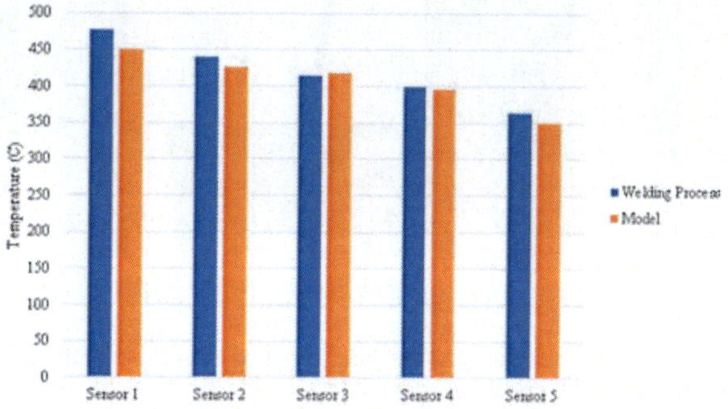

with the experimental data from the thermocouples, which is one indication that the material flow is correctly simulated.

The forces simulated on the tooling also provide a point of validation for the model. Correct forces indicate that the heat generation and material flow are accurate because the flow stresses for AA 2029 at higher temperatures are sensitive to both temperature and strain rate. The forces in the tooling are output by the Bond Technologies RFSSW end-effector. The model predicted the correct patterns and amplitude of the forces experienced by the tooling.

The force on the pin which is shown in Fig. 10 shows the distinct periods of the welding process; from $t = 0$ s to $t = 0.3$ s the pin and shoulder tool are in the clamping phase, both pressing down on the plug material, and the forces exerted on both the pin and the shoulder are increasing linearly; from $t = 0.3$ s to $t = 0.8$ s, the

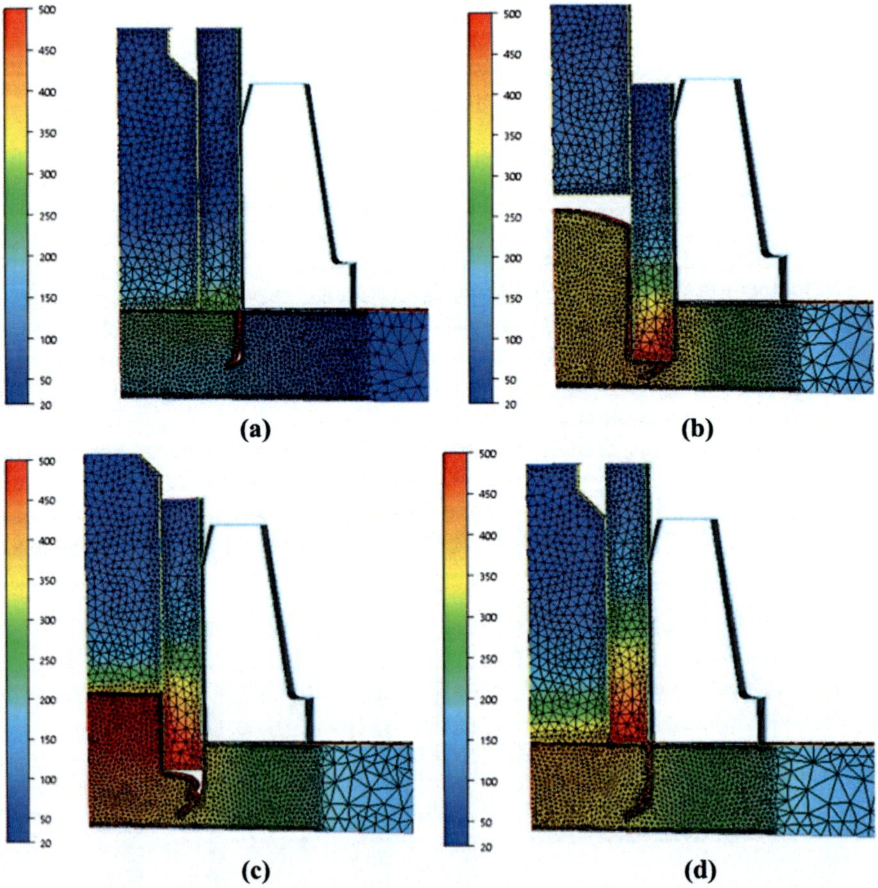

Fig. 9 Visualization of the heat generation in the model: **a** clamping phase, **b** plunge phase, **c** refill phase, **d** final position

Fig. 10 Forces on the pin tool during the RFSSW repair process

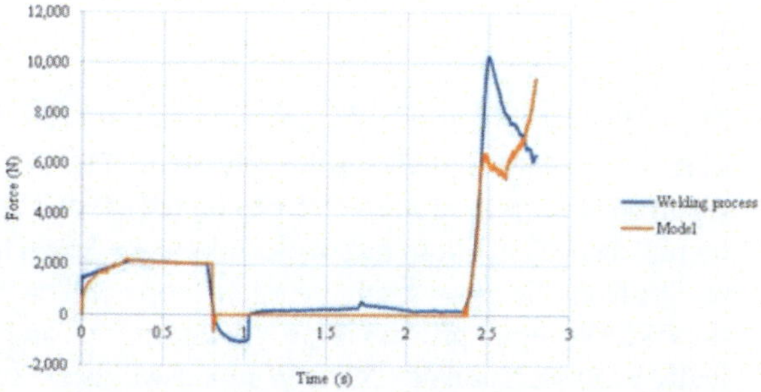

pin and shoulder are still rotating and they are under constant load; from $t = 0.8$ s to $t = 2.3$ s, the pin is rising away from the deformed material and has no contact with the material; from $t = 2.3$ s to $t = 2.8$ s, the pin is refilling the cavity left by the shoulder, pressing down and both the welding process and the model show force peaks of about 10 kN.

Forces in the shoulder tool were very well predicted by the model as shown in Fig. 11. The patterns and the peaks are very similar in this comparison. The same distinct periods of the welding process are visible in this plot as we see the clamping phase of increasing force from $t = 0$ s to $t = 0.3$ s, the constant force period where the tools are both rotating without plunging from $t = 0.3$ s to $t = 0.8$ s, the plunging phase of the shoulder where force is increasing from $t = 0.8$ s to $t = 2.3$ s, and the refill stage where the shoulder evades material contact until the end of the welding cycle as material presses up against the shoulder tool.

Forces and temperatures predicted by the model are similar to those of the welding process, which is a good validation of the simulation work. The material flow predicted by the model, where a bonded interface under the shoulder is visible at the end of welding, is also accurate to the experiment. Future work will focus on a criterion for prediction of bonding, as the current model only indicates where bonding may occur but does not predict whether adequate bonding did occur.

Fig. 11 Forces on the shoulder tool during the RFSSW repair process

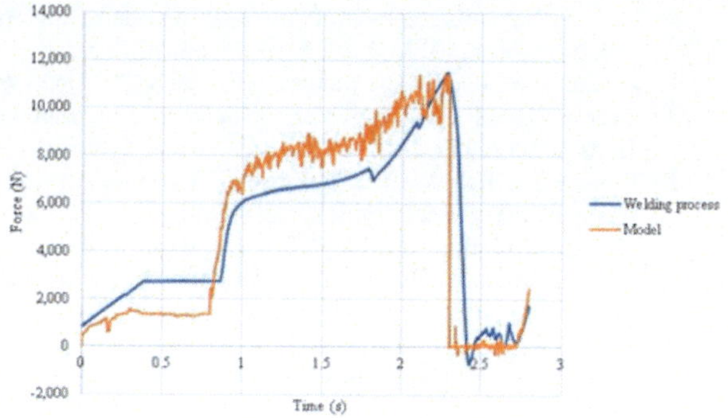

Conclusions

This RFSSW repair process was successfully demonstrated to be a viable method of repair for interrupted and aborted welds. The 2D axisymmetric model of RFSSW repair successfully captured the physics of a repair weld and was validated by experimental data. The following conclusions were drawn from this work: The experimental results from the cross-sectional microscopy and the tensile strength testing show that the RFSSW repair process improves interrupted and aborted spot welds both aesthetically and mechanically. RFSSW repair welds are 6% stronger than typical RFSSW welds.

1. The model successfully predicts the heat generation from the friction between the tools and the AA2029-T8, as evidenced by its temperature predictions against the experiment.
2. The model successfully demonstrates its ability to predict the forces experienced by the pin and the shoulder tools during welding under, thus demonstrating that the material flow and the plastic deformation of the material are being calculated correctly.

The initial results of the RFSSW repair method show promise in being able to fully repair interrupted and aborted RFSSW welds.

References

1. Lakshmi Balasubramaniam G, Boldsaikhan E, Fukada S, Fujimoto M, Kamimuki K (2020) Effects of refill friction stir spot weld spacing and edge margin on mechanical properties of multi-spot-welded panels. J Manuf Mat Process 4:55
2. Lakshmi Balasubramaniam G, Boldsaikhan E, Joseph Rosario GF, Ravichandran SP, Fukada S, Fujimoto M, Kamimuki K (2021) Mechanical properties and failure mechanisms of refill friction stir spot welds. J Manuf Mat Process 5:118
3. Schilling C, dos Santos J (2004) Method and device for joining at least two adjoining work pieces by friction welding
4. Itapu S, Muci-Küchler K (2006) Visualization of material flow in the refill friction stir spot welding process, 0148-7191. SAE Technical Paper
5. Watanabe M, Feng K, Nakamura Y, Kumai S (2011) Growth manner of intermetallic compound layer produced at welding interface of friction stir spot welded aluminum/steel lap joint. Mater Trans 52:953–959
6. Tarasov SY, Rubtsov VE, Kolubaev EA (2014) A proposed diffusion-controlled wear mechanism of alloy steel friction stir welding (FSW) tools used on an aluminum alloy. Wear 318:130–134. https://doi.org/10.1016/j.wear.2014.06.014
7. Wang G, Zhao Y, Hao Y (2018) Friction stir welding of high-strength aerospace aluminum alloy and application in rocket tank manufacturing. J Mater Sci Technol 34:73–91. https://doi.org/10.1016/j.jmst.2017.11.041
8. Reimann M, Goebel J, dos Santos JF (2017) Microstructure and mechanical properties of keyhole repair welds in AA 7075–T651 using refill friction stir spot welding. Mater Des 132:283–294. https://doi.org/10.1016/j.matdes.2017.07.013
9. Ikuta A, North TH, Uematsu Y (2018) Characteristics of keyhole refill process using friction stir spot welding. Weld Int 32:417–426. https://doi.org/10.1080/09507116.2017.1346859

10. Zhou L, Liu D, Nakata K, Tsumura T, Fujii H, Ikeuchi K, Michishita Y, Fujiya Y, Morimoto M (2012) New technique of self-refilling friction stir welding to repair keyhole. Sci Technol Weld Joining 17:649–655. https://doi.org/10.1179/1362171812Y.0000000058
11. MatWeb. Aluminum alloy heat treatment temper designations. Available online
12. Hovanski Y, Curtis A, Michaelis SE, Blackhurst P, Larsen B (2021) Advances in refill spot welding productivity
13. Larsen B, Hovanski Y (2020) Reducing cycle times of refill friction stir spot welding in automotive aluminum alloys
14. Berger E, Miles M, Curtis A, Blackhurst P, Hovanski Y (2022) 2D axisymmetric modeling of refill friction stir spot welding and experimental validation. J Manuf Mat Process 6:89
15. Prymak DJ (2021) A new method of measuring flow stress for improved modeling of friction stir welding M.Sc. Brigham Young University, Provo

Production Evaluation of Refill Friction Stir Spot Welding

Ruth Belnap, Paul Blackhurst, Yuri Hovanski, Andrew Curtis, Josef Cobb, and Heath Misak

Abstract Since the introduction of refill friction stir spot welding (RFSSW) in the early 2000s, numerous evaluations of properties, performance, and prototypes have been documented. RFSSW is a solid-state spot joining technique that is being evaluated as a rivet replacement technology for aerospace structures and fuselage fabrication. A production evaluation of overlapping wing skin assembly was completed and experimentation on stringer-stiffened curvilinear panels is underway. Weld properties were evaluated for spots produced on bare sheet as well as for sections with sealant applied. Furthermore, evaluations of the performance of repaired spots using the RFSSW process are demonstrated. Characterization of the weld interface, heat-affected zone, and complete through-thickness consolidation via computed tomography is presented.

Keywords Spot welding · Aluminum · Aerospace

Introduction

RFSSW was invented in the early 2000s [1] as a derivative of friction stir welding. The process uses a 3-piece tool set to generate frictional heating and pressure in the workpiece, which enables material flow and bonding of two sheets together. Figure 1 illustrates the steps of the refill spot weld process. A three-piece tool set, shown in Fig. 2a, can vary in size and material [2], allowing this process to work in many different scenarios. In recent years, much research has been conducted to further the

R. Belnap · P. Blackhurst · Y. Hovanski (✉) · A. Curtis
Brigham Young University, 265 CTB, Provo, UT, USA
e-mail: yuri.hovanski@byu.edu

J. Cobb
NASA-MSFC, Huntsville, AL, USA
e-mail: josef.b.cobb@nasa.gov

H. Misak
Spirit AeroSystemsInc, Wichita, KS, USA
e-mail: heath.e.misak@spiritaero.com

Y. Hovanski et al. (eds.), *Friction Stir Welding and Processing XII*,
The Minerals, Metals & Materials Series,
https://doi.org/10.1007/978-3-031-22661-8_13

use of RFSSW in the construction of thin-walled stringer-stiffened panels for the aerospace industry [3–6].

RFSSW is unique as compared to other traditional joining methods—namely riveting, resistance spot welding, and arc welding techniques—because it is a non-additive process that occurs entirely in the solid state. Traditional riveting techniques

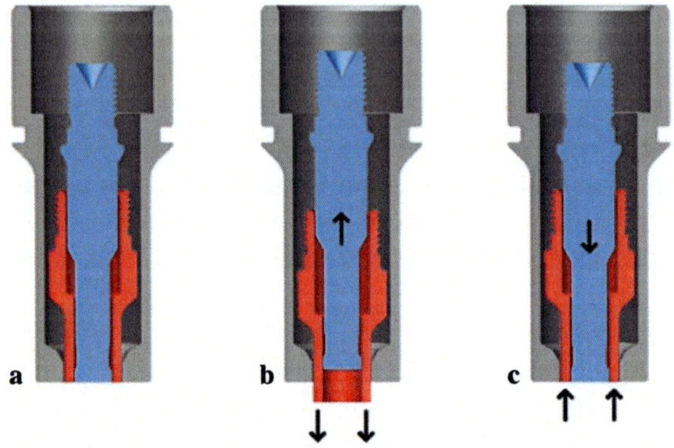

Fig. 1 Schematic representation of the refill friction stir spot welding process

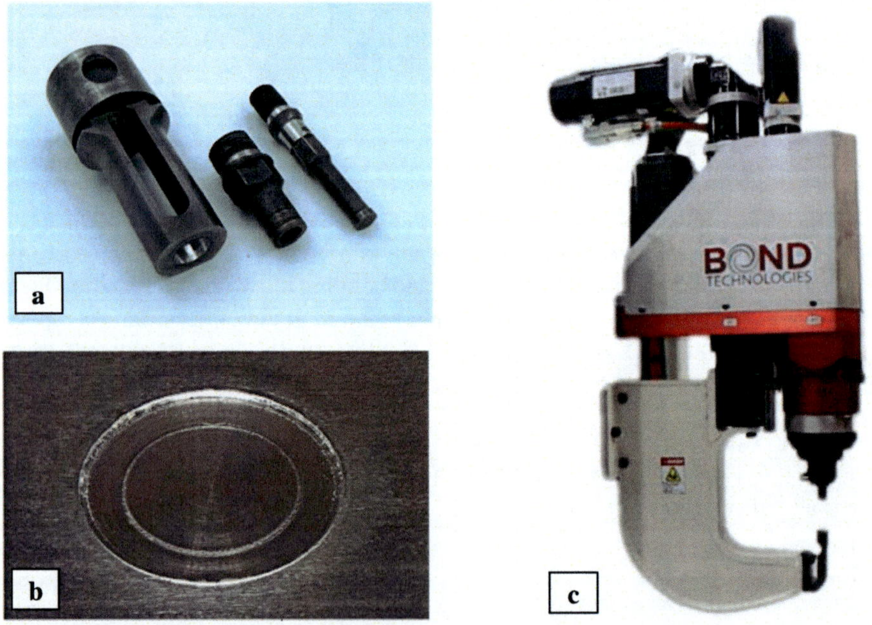

Fig. 2 **a** RFSSW tool set consisting of clamp, shoulder, and probe (from left to right). **b** Refill friction stir spot weld in aluminum sheet. **c** Bond Technologies refill friction stir spot welding end effector

require a hole to be drilled, which can become a crack initiation point. Resistant spot welding and arc welding involve melting, which results in a weak microstructure in the weld nugget and heat-affected zone. In short, RFSSW has the potential to be a strong alternative joining method.

The projects discussed in this paper are expected to illuminate how RFSSW can be utilized in the construction of wing and skin-stiffened structures. This work can be divided into two: the joining of sheet to itself in flat panels, and the joining of stringers to curvilinear panels. Both configurations consisted of joining 2xxx to 2xxx series aluminum; further, specifics will not be discussed herein.

Development for both configurations follows the same general approach. It starts with establishing the process parameters for single-spot welds and continues to multi-spot development and small-scale prototypes. Then, the final panel is produced and prepared to be sent away for large-scale testing at another facility. Note that as of the time of writing, the flat panels are completed, and the curvilinear panels are in the small-scale testing phase of development.

Methodology and Equipment

This work was accomplished with a position-controlled refill friction stir spot welder supplied by BOND Technologies, shown in Fig. 2c. This machine can produce weld joints at sub-one-second cycle times [7], and it is able to collect force and kinematic data for the clamp, the shoulder, and the probe. This data proved to be crucial in developing strong and reliable process parameters for the material sets in this study.

Tooling—The tooling was also sourced from BOND Technologies. An H13 tool was used for the development and production of the flat panels. Due to issues that are discussed later in this paper, the stringers and curvilinear panels were produced using a tungsten carbide tool set with the identical geometry.

Single-Spot Development

The first task of this work was to define a parameter set for a single-spot weld and validate that its mechanical properties are within a satisfactory range. As mentioned earlier, this phase of development was repeated to refine two parameter sets for the two material sets used in this project. Establishing a reliable single-spot weld was important for analyzing results later in the study.

Specimen geometry—The development begins with preparing one-by-three-inch sheets of aluminum to weld together in single-spot coupons. This size and configuration work well for tensile testing to validate strengths, as well as CT scanning and optical microscopy to check for voids.

Material preparation—Prior to welding, each specimen is prepared identically. They are cleaned thoroughly with isopropyl alcohol and then sanded with 80 grit sandpaper on a die grinder to remove the oxide layer. This preparation occurs within 24 h of welding.

Tensile testing—Tensile testing is one of the primary methods used to verify the quality of an RFSSW weld. After letting the welds age for at least 96 h, they are pulled on an Instron tensile tester; the data collected from this test includes the ultimate lap shear strength as well as the fracture mode.

Analyzing fracture modes is useful because it provides insight into weld strength. When the weld nugget pulls out of the top coupon completely, as seen in Fig. 3a, it signifies that the nugget is stronger than the base material, which is highly desirable.

Optical microscopy—During single-spot development, it is common to reserve one weld of each set to be cross-sectioned, polished, and inspected under an optical microscope. The microscopy reveals whether the welds are fully consolidated or if they have voids, which can indicate that the kinematics of the weld need to be adjusted. It also verifies that the surface finish is acceptable. Figure 3b shows a microscope image of a successfully consolidated and flat weld.

CT scanning—Computed tomography scanning is another valuable tool for identifying voids in a weld. A CT scan will show the weld in three dimensions so that

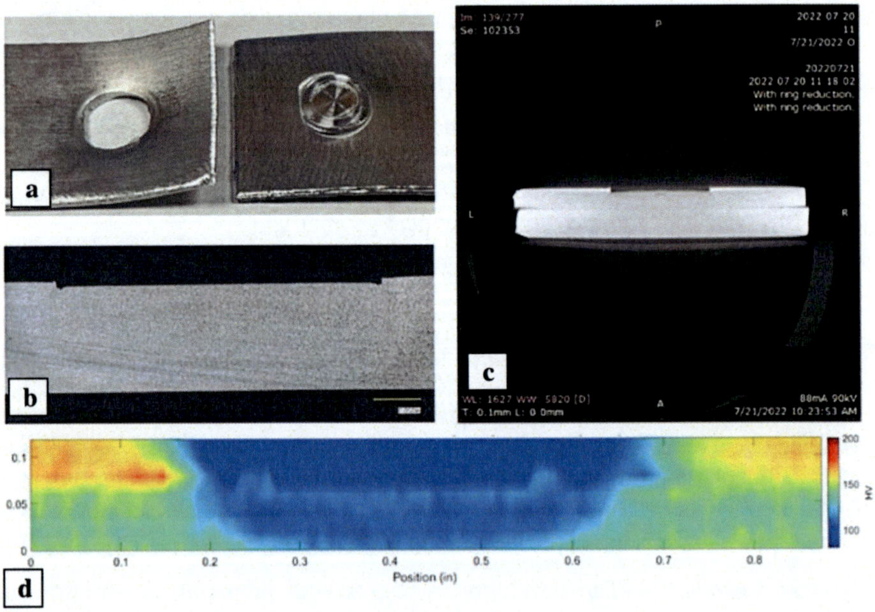

Fig. 3 **a** Nugget pullout fracture mode of an RFSSW after a tensile test. **b** Optical microscopy of the cross section of a consolidated RFSSW. **c** Computed tomography scan of an RFSSW. **d** Microhardness map of the cross section of an RFSSW

an entire weld, rather than just a single plane, can be checked for consolidation. Figure 3c shows what a CT scan of an RFSSW weld looks like.

Microhardness—Microhardness testing is commonly employed in this phase of development to validate strength. It provides high-resolution data about a weld cross section that indicates whether a weld is getting sufficient heat. Figure 3d is an example of a microhardness map of a weld.

Single-spot results—After iterating through these steps several times, a parameter set was chosen for both material sets. These parameter sets created welds that were validated as reliably strong, had void-free cross sections, and displayed appropriately flat surface finishes.

Flat Panel Production

After narrowing in on a parameter set, the next task was producing flat panels. This involved welding together two large flat panels with a three-inch overlap. New concerns came under review at this juncture, namely fixturing, distortion, sealant application, weld spacing and indexing, and weld order.

Flat panel geometry—The panels are made of two 36″ by 36″ sheets spot welded together on a three-inch overlap. There are 35 rows and three columns of spots spanning the overlap, for a total of 105 welds. Three of these panels were completed.

Flat panel fixture—Fig. 4 shows the fixture setup that was designed to hold the large panels. It was constructed out of 80/20 aluminum extrusion and 3D printed parts.

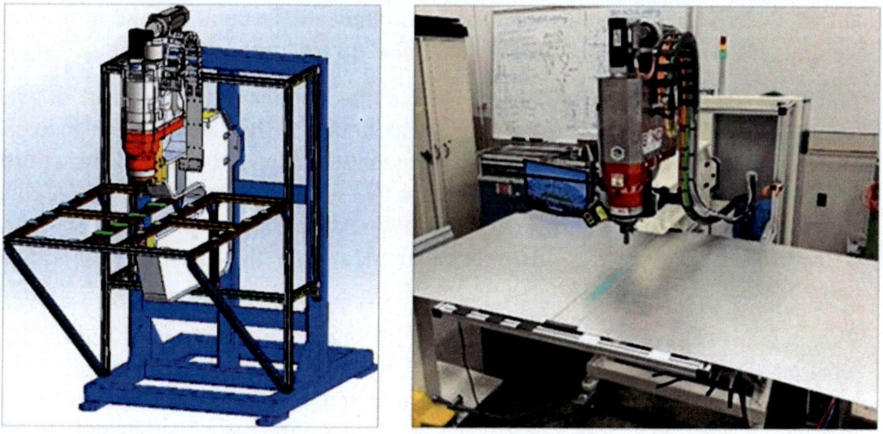

Fig. 4 **a** CAD model of fixture structure to hold the flat panels. **b** Actual flat panel setup on RFSSW machine

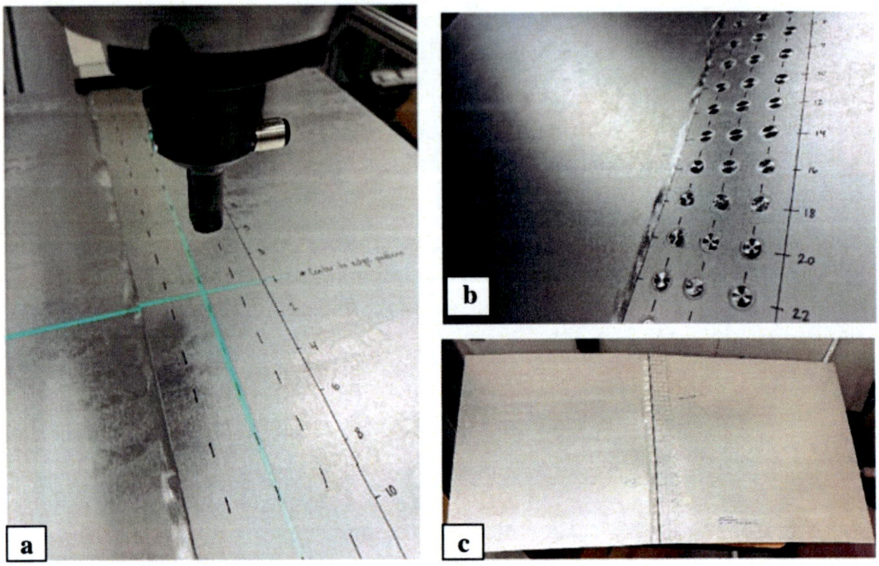

Fig. 5 **a** Closeup of the indexing marks on the panels. **b** Closeup of post-weld indexing. **c** The third completed flat panel

Weld order—The weld spacings and order were calculated by an RFSSW ANSYS computer model, and they were optimized to minimize distortion. To validate the results from the computer, several smaller prototypes were welded with different weld orders; it was concluded that the weld order from the computer model did indeed minimize distortion, and welding proceeded with that spacing and order.

Indexing—Prior to welding, the panels were marked up with dotted lines, as seen in Fig. 5a, and a laser level was calibrated to indicate the center of the spot. During welding, the panel was held in place and lined up for each weld manually.

Tool material change—Over the course of this production, a handful of welds failed. These occurred because as the temperature of the steel tool set increased, the aluminum became more likely to stick to the steel than to itself, causing a portion of aluminum to come out of the weld with the tool set and leave a cavity. These failures indicated that a change to tungsten carbide would be necessary for the next section of the project. The failed welds were repaired, and the flat panels were completed with the steel tool set.

Weld repair—The weld failures were not so severe that they could not be easily repaired. This was accomplished by welding a spot directly over the failed weld. That process resulted in completely consolidated welds that were of comparable strength to joints that needed no repair. Details about the development and results of this repair were documented previously [8].

Sealant—Previous experiments concluded that, when fully consolidated into the weld, sealant contributes to a decrease in ultimate lap shear strength. Therefore, a sealant study was conducted in order to find the maximum sealant bead thickness possible that would still prevent sealant from leaking into the welds.

As part of this study, three 9-spot coupons were prepared, and each was precisely applied with beads of sealant around the outer perimeter. The three coupons corresponded with three different thicknesses of sealant beads. Optical microscopy was performed on all 27 welds, and there was no evidence of sealant leaking into any of the welds. Therefore, it was concluded that a 1/8″ bead of sealant would be an acceptable thickness and was applied around the perimeter of the flat panel overlap.

Flat panel results—The production of these flat panels, two without sealant and one with sealant, has been successfully completed. The third panel, with sealant, is shown in Fig. 5c. The welded panels are currently undergoing ultrasonic and tensile testing, the results of which are outside of the scope of this paper. Each of the 105 welds had a good surface finish and the overall distortion was minimal.

Curvilinear Panel and Stringers

After successfully welding the flat panels, the subsequent assignment was to weld stiffening stringers onto curvilinear panels. As mentioned previously, development and production were executed using a tungsten carbide tool set. Single-spot weld development occurred for this new material set and a parameter set was defined. As this project is ongoing, this section of the paper will address the preliminary problems of fixturing the panel, fixturing the stringers, and indexing.

Curvilinear panel geometry—There are two versions of these panels that will be produced. One is a sheet 28″ by 80″ that will have two stringers welded onto it, and the other is a 37″ by 80″ and will have three stringers welded onto it. There will be a total of five panels produced: two with two stringers and three with three stringers. Both designs have a curvature with a 76.5″ radius.

Curvilinear panel fixture—The fixturing for this setup built off of the fixture initially created for the flat panels, but the larger curvilinear panels required additional external support. Furthermore, it was also necessary to design stanchions and supports with the same radius as the panels. The curved panel fixture is pictured in Fig. 6a and b.

Stringer fixture—One issue that came up was that the location of the spot weld caused the edge of the stringer to bulge out during welding which can be seen in Fig. 7c. This was a concern because that material displacement presents an opportunity for the weld to not fully consolidate. As such, an additional fixture, which can be seen in Fig. 7a and b, was designed to fix the bulging. The fixture leaves a slot for the end effector to fit in and perform the welds, and it clamps the edge and successfully prevents it from blowing out, shown in Fig. 7d.

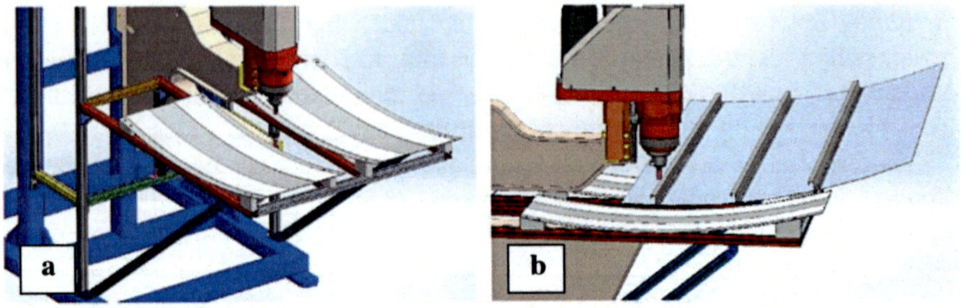

Fig. 6 **a** CAD model of curved panel fixture. **b** CAD model of fixture with the stringer-stiffened panels on it

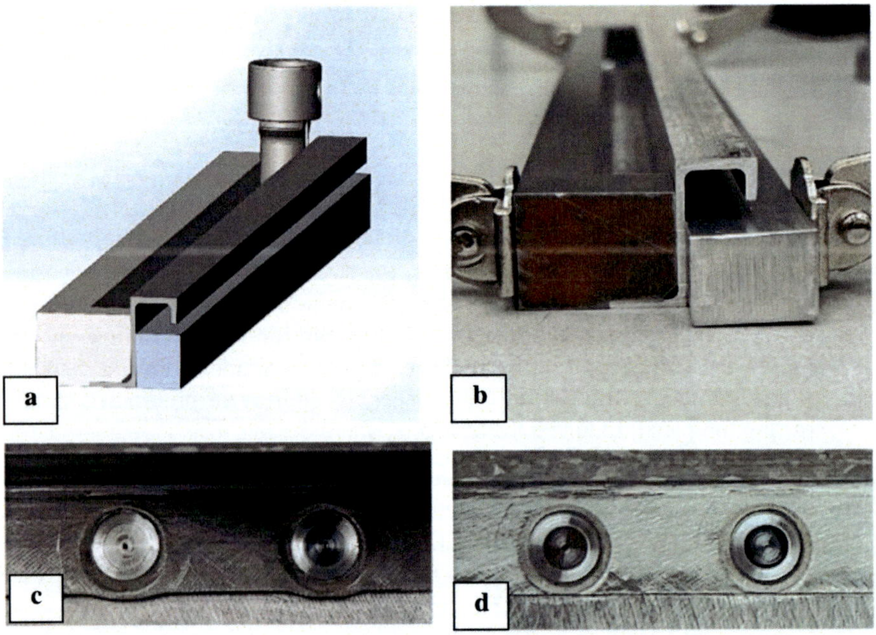

Fig. 7 **a** CAD model of a the stringer, fixture and clamp from the tool set. **b** Closeup of the fixture clamped onto a stringer section. **c** Welds performed without the fixture with noticeable blowout on the edge. **d** Welds performed with the fixture with no blowout

Indexing—The clamp fits into the channel between the stringer and its fixture and is indexed against the stringer, which locates the spot at the edge of the stringer. In the longer dimension, the spots are marked at one-inch intervals and aligned manually.

Future Work

Curved panel completion—At this point in the project, test pieces have been completed to validate that the fixtures and indexing methods work well. The 80-inch stringers will be welded to the five planned curvilinear panels in the coming months.

Fatigue study—Plans for collecting fatigue data for the material set used in the stringer-stiffened panel are underway.

Sealant on stringers—There will also be an exploration into the application of sealant on stringers. It is not anticipated that the method used for the flat panels will transfer well to the stringers, so this problem will require its own development.

Clamp modifications—In using this toolset for the duration of this study, some concerns have come to light regarding the clamp design, and there are proposed modifications to reconcile these concerns. These include adding a chamfer to its inner diameter, adding a radius to its outer diameter, and creating a mounting hole for a thermocouple.

Conclusions

In this study, RFSSW has been demonstrated to be a suitable replacement for traditional riveting and joining techniques in the production of wing and skin-stiffened structures. Refill welds are less susceptible to corrosion due to the hermitic nature of the joint. They also have exhibited good rigidity in addition to high fatigue strength and high ultimate lap shear strength. Furthermore, because these welds consist of fewer parts than rivets do, they also can be cheaper and lighter joints.

The conclusions of this study are summarized below:

- RFSSW is a suitable joining method for the production of wing spar constructions and skin-stiffened structures.
- Sealant applied around the perimeter of the weld area provides adequate sealing and does not affect weld quality.
- Repairs can be successfully performed without reduction in original mechanical properties.
- Panel distortion can be minimized with weld order.
- Stringer blowout can be counteracted with appropriate fixturing.

References

1. Christoph S, dos Santos J (2004) Method and device for joining at least two adjoining work pieces by friction welding. In: Office UP (ed) United States, pp 1–8
2. Shen Z, Ding Y, Gopkalo O, Diak B, Gerlich AP (2018) Effects of tool design on the microstructure and mechanical properties of refill friction stir spot welding of dissimilar Al alloys. J Mater Process Technol 252:751–759
3. Kubit A, Trzepiecinski T, Swiech L, Faes K, Slota J (2019) Experimental and numerical investigations of thin-walled stringer-stiffened panels welded with RFSSW technology under uniaxial compression. Mater 12(11)
4. Kubit A, Bucior M, Kluz R, Bochnowski W, Perlowski R (2019) Experimental research of the weakening of the fuselage skin by RFSSW single row joints. Adv Sci Technol Res J 13(3):90–97
5. Kubit A, Trzepiecinski T, Faes K, Drabczyk M, Bochnowski W, Korzeniowski M (2019) Analysis of the effect of structural defects on the fatigue strength of RFSSW joints using C-scan scanning acoustic microscopy and SEM. Fatigue Fract Eng Mater Struct 42(6):1308–1321
6. Kubit A, Jurczak W, Trzepiecinski T, Faes K (2020) Experimental and numerical investigation of impact resistance of riveted and RFSSW stringer-stiffened panels in blunt impact tests. Trans FAMENA 44(3):47–58
7. Larsen B, Hovanski Y (2020) Reducing cycle times of refill friction stir spot welding in automotive aluminum alloys. SAE Technical Paper 2020-01–0224. https://doi.org/10.4271/2020-01-0224
8. Berger BPE, Belnap R, Miles M, Hovanski Y (2023) Refill friction stir spot welding repair – simulation and validation. In: Hovanski Y, Sato Y, Upadhyay P, Naumov A, Kumar N (eds) Friction stir welding and processing XII. Springer, Cham

Embedded Anchoring of Multi-Material Assemblies by Friction-Riveting Process

Hrishikesh Das, Keerti S. Kappagantula, Abhinav Srivastava, Piyush Upadhyay, Jorge F. Dos Santos, and Md. Reza-E-Rabby

Abstract In this paper, we report on work that focused on extending the capability and broaden the applicability of the friction-riveting process for joining stacks composed of a wide range of multi-layer polymer to metals and similar and dissimilar metals. We first present direct experimental evidence of our use of this process to join aluminum-to-steel dissimilar metals with steel rivets. We have demonstrated for the first time the use of magnesium as a rivet material for joining carbon fiber reinforced polymer (CFRP) stacks, aluminum as a rivet material to laminate magnesium to CFRP, and aluminum-to-aluminum similar metal joining and aluminum-to-steel dissimilar metal joining with steel rivets. Our work sheds light on detailed process parameter optimization and the corresponding process response behavior, thus advancing our understanding of this complex joining method for a wide range of material combinations.

Keywords Friction-riveting · Magnesium rivet · Carbon fiber reinforced polymer · Dissimilar metal

Introduction

An important goal in global markets is to curtail carbon emissions and fuel consumption while maintaining material strength-to-weight ratios, leading to research toward multi-material concepts for assembly of products during manufacturing processes. Stacking of multi-materials, including different grades of polymers with metals, is an area of high interest in the aircraft and automotive industries [1, 2]. In recent years, various aircraft components, such as fuselages and wings, are being fabricated using different grades of polymer composites or polymer–metal joints. Conventional joining methods are not appropriate for such complex stacked components. Although

H. Das · K. S. Kappagantula · A. Srivastava · P. Upadhyay · J. F. Dos Santos ·
Md. Reza-E-Rabby (✉)
Applied Materials and Manufacturing Group, Pacific Northwest National Laboratory, Richland, WA, USA
e-mail: md.reza-e-rabby@pnnl.gov

Y. Hovanski et al. (eds.), *Friction Stir Welding and Processing XII*,
The Minerals, Metals & Materials Series,
https://doi.org/10.1007/978-3-031-22661-8_14

joining techniques such as adhesive bonding [3, 4] and fastening methods [5, 6] are used for these types of material stacks, several shortcomings still exist. Some limitations associated with adhesive bonding are 1) temperature sensitivity, 2) hazardous chemical handling, 3) curing methods that limit assembly line speed, and 4) unpredictable fracture properties under different conditions. Similarly, limitations with fastening methods include 1) loosening of fasteners due to vibration, creep, and temperature variation; 2) the need to access to both sides of parts to be joined, and 3) stress concentration.

Friction-riveting (Fric-Riveting), a new joining process invented by Helmholtz-Zentrum Geesthacht [7, 8], has demonstrated the capability to alleviate the issues mentioned above. Recently, Santos et al. have reported on demonstrations in which several sets of polymers and metals were joined using Fric-Riveting [9–18]. It is intriguing that in previous studies [9–18], different grades of titanium (Ti-6Al-4V, Ti grade 2) and aluminum (Al) rivets were inserted in thick polymers or stacks of polymers (CF-PEEK, Polyethermide, GFRP, PEI-GF, Polymide-6, etc.), whereas only Ti was used for polymer and Al stacking [14, 16]. However, to the best of our knowledge, there are no published reports of magnesium (Mg) being used as a rivet material to stack CFRPs or for similar and dissimilar metal joining by Fric-Riveting.

At Pacific Northwest National Laboratory, we are working with a leading industrial manufacturer to address problems associated with joining similar and dissimilar materials. This study reports on the first experimental demonstration of Case I-AZ31-Mg as a rivet material for CFRP stacking, Case II-AA6061-T6 as a rivet material for stacking AZ31 Mg–CFRP (with the Mg on top), Case III-cast A356 to dual-phase (DP) 590 steel joining with an M42 rivet, and Case IV-threaded low-carbon-steel rivet for stacking AA6061-T6–AA6061-T6 by Fric-Riveting process. Representative microstructures and corresponding process response variables were analyzed and described in detail for different materials and joint configurations.

Experimental Methods

We used four CFRP laminates (3 mm thick each) as a stack material for Case I. This CFRP laminates are commercially known as Ultramid Advanced N XA-3454, manufactured by BASF corporation. PA9T thermoplastic polymer is reinforced with short carbon fiber (40 vol%) to form this composite. For the Case II configuration in this study AZ31 Mg (2.4 mm thick, UTS: 260 MPa, Hardness: 65) sheets were laminated with CFRP (3 mm thick each). Standard AZ31 Mg alloy, AA6061-T6 (UTS: 310 MPa, Hardness: 107 Hv) and AA7075-T6 (UTS: 572 MPa, Hardness: 175 HV) were used as rivet materials (diameter 5 mm) for Case I and II, respectively. Design details of rivets for Cases I and II are shown with Fig. 1. A356 cast aluminum (4 mm thick, UTS: 200 MPa) and DP 590 steel (1 mm thick, UTS: 590 MPa) were used as a stack in Case III. In Case IV, stacking materials were standard AA6061-T6 (3 mm thick) to AA6061-T6 (3 mm thick). Hardened M42 tool steel (diameter 6.35 mm, hardness: 65RC) and high-strength steel threaded rod (diameter 6.35 mm, hardness:

33RC) were used as rivet materials for Cases III and IV, respectively. Respective process parameters (tool rotational speed [RPM], plunge speed, and forge force) for 4 set of stacking listed in Table 1 and are further explained with Fig. 1c, Fig. 2b, 3b, and 4b, respectively.

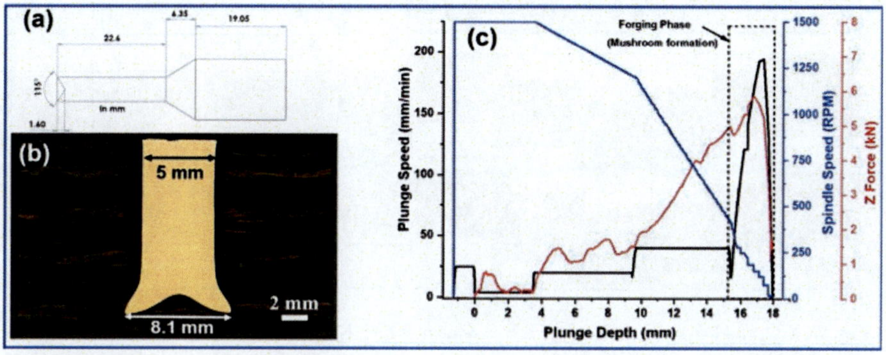

Fig. 1 **a** Rivet design. **b** Macro-image of the AZ31 Mg rivet insertion through CFRP stacking. **c** Process response

Table 1 Material stacking and process parameters (RPM, plunge speed, max. forge force)

Approach	Material stack	Rivet	RPM	Plunge speed (mm/min)	Max. forge force (kN)
I	4 CFRP sheets of 3 mm each	5 mm AZ31	**1500**	**4**	5.8
			1200	**20**	
			400	*40*	
			20	*780*	
II	3 CFRP sheets of 3 mm each – 2.4 mm AZ31	5 mm AA6061-T6	**1200**	**4**	4.5
			1200	**20**	
			1200	*20*	
			1200	*50*	
III	A356 cast Aluminum – Dual Phase (DP) 590 steel	6.35 mm M42 steel	**1500**	**4**	27
			1500	**20**	
			1500	*20*	
			1500	*50*	
IV	3 mm 6061(T6)	High strength steel threaded rivet	**1500**	**4**	10
			1200	**20**	
			1000	*20*	
			800	*25*	

(**Bold**: drilling phase, *Italic*: forging phase)

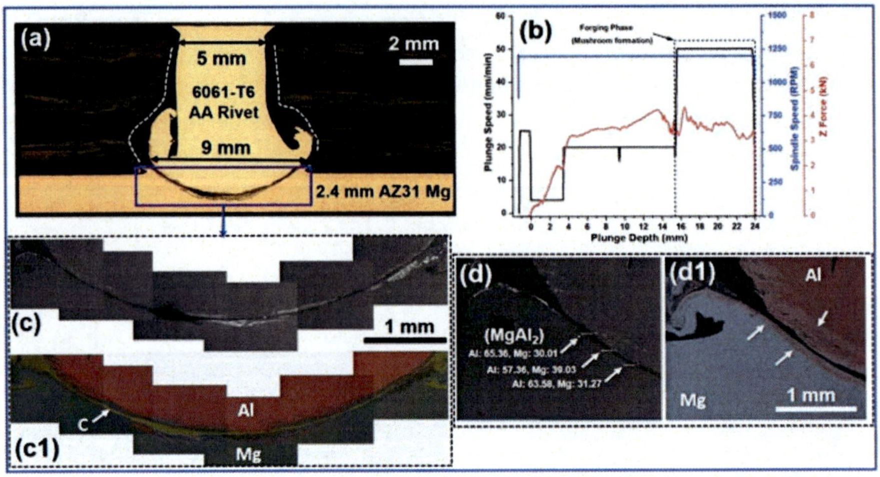

Fig. 2 **a** Optical macrograph of CFRP–2.4 mm AZ31 Mg sheet stacking with an AA 6061-T6 rivet. **b** Process response variables. **c** SEM image and corresponding **c1** EDS elemental mapping at the interface of AZ31 Mg sheet and AA6061 rivet. **d** SEM and **d1** EDS elemental mapping with point analysis at the interface near the left hook

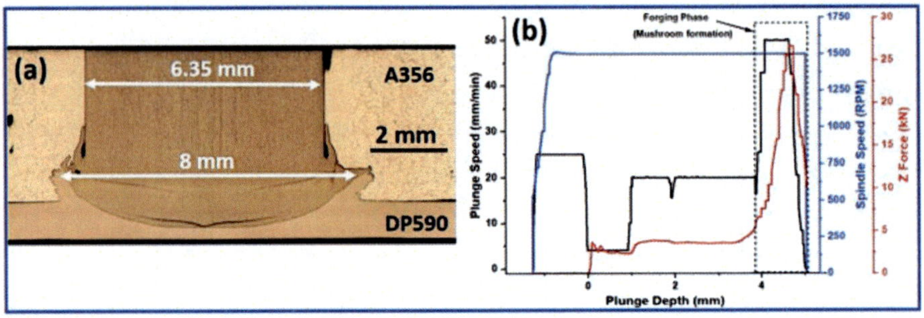

Fig. 3 **a** Optical macrograph of 4 mm A356–1 mm DP590 stacking with a M42 steel rivet. **b** Process response

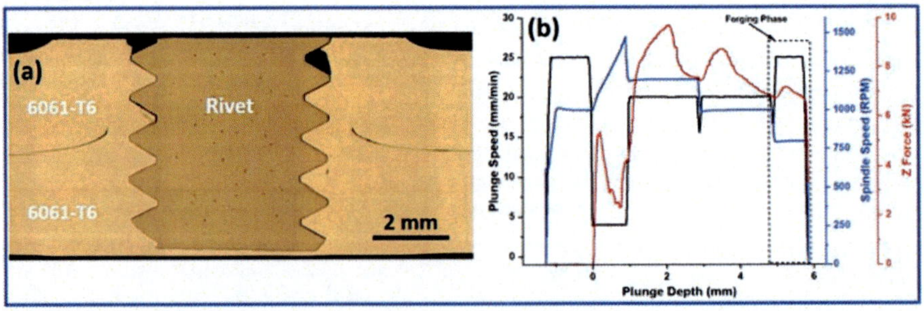

Fig. 4 **a** Optical macrograph of a 3 mm AA6061–3 mm AA6061 stack with low strength threaded steel rivets. **b** Process response

Results and Discussion

Case I: 4 CFRP Sheet Stacking with AZ31 Mg Rivet

A typical macrograph of direct friction-riveted joints is presented in Fig. 1b. The optimum process parameters were achieved after multiple trials. The scheme presented in Fig. 1c is divided into two phases of friction-riveting: 1) drilling and 2) forging. The first two plunge steps (Table 1, highlighted in **bold**) are drilling of the rivet through the CFRP. During the drilling phase, high heat generation is required to facilitate the softening of the top surface of the stack; therefore, a higher rivet RPM and a lower plunge speed is selected, thus allowing the rivet to penetrate the sheets easily. The second phase is essentially the forging step (Table 1, highlighted in *italic*). Here, the RPM is lower with very fast plunge speed. Because the rivet is already heated during the previous drilling phase, a fast plunge speed of 780 mm/min at this step enables plastic deformation of the rivet tip, which also restricted by the lower CFRP sheet resulting in mushrooming or flaring at the rivet tip (Fig. 1c) within the sheet. The formation of mushrooming is further supported by the sudden increment in Z force (marked with black dotted box, Fig. 1d) toward the end of plunge depth that induces hydrostatic pressure on the rivet. It is important to note that at the high forge force toward the end of the process, the plunge depth reads as 18 mm in Fig. 1d, but in the optical micrograph, this depth is 12.9 mm (Fig. 1c). This difference is due to the mushrooming effect as the initial rivet tip diameter of 5 mm deformed and flared up to 8.1 mm (Fig. 1c).

Case II: 3 CFRP Sheet to 2.4 mm AZ31 Mg Stacking with an AA6061-T6 Rivet

A typical macrograph of the CFRP–2.4 mm AZ31 stack with AA6061-T6 rivets is shown in Fig. 2a. It is interesting to note that after a certain plunge depth, the Z force increases during the rest of the process (Fig. 2b). This is obvious as the rivet first is fed through the CFRP stack and then approaches the Mg sheet. It is evident from the macrograph that during the joining process, the Al rivet first engages in the Mg sheet, followed by plastic deformation at the tip. The resistance to penetration through the Mg sheet is much higher than that at Case I (Fig. 1b), and the resulting metallic deformational flow in both the rivet and the Mg sheet is noticeable here with dominant mushrooming of the rivet as flow stress decreases. The difference between the commanded plunge depth (Fig. 2b as 24 mm) and the rivet length measured by the optical microscope (9.8 mm in Fig. 2a) is due to the mushrooming effect as the initial rivet tip diameter of 5 mm deformed to 9 mm (Fig. 2a). Bonding between the polymer and the metal also is enhanced by mechanical interlocking observed at the interface between CFRP and AZ31. The rotational speed was kept constant at 1200 RPM throughout the process as the Al rivet is fed through the Mg to accomplish bonding. This also caused a certain amount of fiber consolidation near the rivet as marked with white dotted lines in Fig. 2a.

Scanning electron microscopy (SEM) and energy dispersive X-ray spectroscopy (EDS) elemental mapping (Fig. 2c and c1) exhibit a vortex flow pattern near the interface of both the Al and Mg sides, indicating formation of brittle Al–Mg Intermetallics (IMCs). To confirm this phenomenon, elemental point analysis (Fig. 2d and d1) and mapping were performed near the left side hook close to the interface. Elemental spot analysis corroborated the probable formation of a $MgAl_2$ IMCs in the Mg side close to the interface (elemental wt% corresponding to each point shown in Fig. 2d). Also, some extent of Al and Mg diffusion was detected near the interface; however, no prominent continuously bonded brittle IMCs layer was found at the interface.

Case III: 4 mm A356 Cast Aluminum to 1 mm DP590 Steel with M42 Steel Rivet

In Case III, we successfully joined dissimilar materials composed of 4 mm A356 cast aluminum alloy to 1 mm DP 590 steel sheets using Fric-Riveting. A typical macrograph of A356–DP590 stacking with M42 steel rivets is shown in Fig. 3a. Multiple trials with various rivet materials and process parameters led to an optimized process parameter combination. The process response behavior is shown in Fig. 3b. The Z force remains constant during the first phase of the riveting process; toward the end of the process, a sudden spike in the Z force is observed, which is the forging phase leading to mushroom formation at the rivet tip. Unlike Cases I and II, a constant high RPM (spindle speed) is maintained throughout the process to generate enough heat to facilitate the rivet drilling through 4 mm cast Al. Increment in plunge speed during the forging phase caused the plastic deformation and mushroom formation at the rivet tip. Intriguingly, for this case, a maximum forge force of 27 kN was observed, which is much higher compared to other three cases. This is due to the high hydrostatic pressure required to reduce flow stress of the steel rivet at this operational temperature to facilitate mushroom formation.

The very small difference between the commanded plunge depth from the process data and the measured rivet length seen in the optical micrograph is due to the mushrooming effect as the initial rivet tip diameter of 6.35 mm deformed to 8 mm (Fig. 3a). Moreover, due to higher resistance from the DP 590 steel, the rivet only was able to penetrate half of its thickness (~0.5 mm). The riveting process also would be able to break the dendritic structure near the outside rivet wall.

Case IV: 3 mm AA 6061 to 3 mm AA6061 with Threaded Rivet

In Case IV, we successfully joined two AA6061-T6 (3 mm thick each) sheets using Fric-Riveting. A typical macrograph of the stack is shown in Fig. 4a. The process response behavior is shown in Fig. 4b. A local maximum Z force was observed throughout the riveting process. Unlike Case III, rivet tip deformation was not observed in this case. This is attributed to the fact that the sheets are Al alloys that deform more easily than steel rivets; hence, the hydrostatic pressure on the rivet is not sufficient for mushroom formation. However, the presence of threads on the rivet provides more contact surface area, resulting in sufficient mechanical joining to form a sound weld. Note that the sheets are welded together near the rivet, which enhances the joint performance significantly. As no mushroom formation occurred in this case, the end plunge depth (6 mm) (Fig. 4b) is similar to the actual depth of penetration measured optically (Fig. 4a). The selection of spindle speed at various stages was based on the fact that only metals are involved in this assembly; hence, high rpm was used during drilling. Once the sheet is heated up, the drill RPMs are decreased, which assists in deformation and rivet penetration.

Conclusions

We have successfully demonstrated four different approaches: Case I, AZ31-Mg as a rivet material for CFRP stacking; Case II, AA6061-T6 as a rivet material to a AZ31 Mg–CFRP stack (Mg on top), Case III, cast A356–DP 590 steel joint with an M42 rivet, and Case IV, threaded low-carbon-steel rivet to stack AA6061-T6–AA6061-T6 by Fric-Riveting. The selections of rivet materials and rivet designs performed efficiently for all the combinations in terms of forming mushroom and/or interlocking features. Process response is a good indication of rivet tip deformation and mushroom formation.

References

1. Borba NZ, Blaga L, dos Santos JF, Amancio-Filho ST (2018) Direct-friction riveting of polymer composite laminates for aircraft applications. Mater Lett 215:31–34. Retrieved from https://www.sciencedirect.com/science/article/pii/S0167577X17317913
2. Borba NZ, Körbelinb J, Fiedler B, dos Santos JF, Amancio-Filhoc ST (2020) Low-velocity impact response of friction riveted joints for aircraft application. Mater Des 186:108369. Retrieved from https://www.sciencedirect.com/science/article/pii/S026412751930807X
3. Stokes VK (1989) Joining methods for plastics and plastic composites: an overview. Polym Eng Sci 29:1310–1324
4. Maguire DM (1989) Joining thermoplastic composites. SAMPE J 25:11–14
5. Rudolf R, Mitschang P, Neitzel M, Rueckert C (1999) Welding of high-performance thermoplastic composites. Polym Compos 7:309–315

6. Yousefpour A, Hojjati M, Immarigeon JP (2004) Fusion bonding/welding of thermoplastic composites. J Thermoplast Compos Mater 17:303–341
7. Amancio-Filho ST, Beyer M, dos Santos JF (2011) Verfahren zum Verbindeneines metallischen Bolzens miteinem Kunststoff-Werkstück. In: European patent no. EP1790462B1
8. Amancio-Filho ST, Beyer M, dos Santos JF (2009) Method of connecting a metallic bolt to a plastic workpiece. In: United States patent no. US7575149B2. Retrieved from https://patents.google.com/patent/US7575149B2/en
9. Cipriano GP, Ahiya A, dos Santos JF et al (2020) Single-phase friction riveting: metallic rivet deformation, temperature evolution, and joint mechanical performance. Weld World 64:47–58. https://doi.org/10.1007/s40194-019-00803-3
10. Borba NZ, Kötter B, Fiedler B, dos Santos JF, Amancio-Filho ST (2020) Mechanical integrity of friction-riveted joints for aircraft applications. Compos Struct 232:111542. https://doi.org/10.1016/j.compstruct.2019.111542
11. Blaga L, Bancilă R, dos Santos JF, Amancio-Filho ST (2013) Friction riveting of glass–fibre-reinforced polyetherimide composite and titanium grade 2 hybrid joints. Mater Des 50:825–829. https://doi.org/10.1016/j.matdes.2013.03.061
12. Altmeyer J, dos Santos JF, Amancio-Filho ST (2014) Effect of the friction riveting process parameters on the joint formation and performance of Ti alloy/short-fibre reinforced polyether ether ketone joints. Mater Des 60:164–176. https://doi.org/10.1016/j.matdes.2014.03.042
13. Borges MF, Amancio-Filho ST, dos Santos JF, Strohaecker TR, Mazzaferro JAE (2012) Development of computational models to predict the mechanical behavior of friction riveting joints. Comput Mater Sci 54:7–15. https://doi.org/10.1016/j.commatsci.2011.10.031
14. Blaga L, dos Santos JF, Bancila R, Amancio-Filho ST (2015) Friction riveting (fric-riveting) as a new joining technique in GFRP lightweight bridge construction. Constr Build Mater 80:167–179. https://doi.org/10.1016/j.conbuildmat.2015.01.001
15. Cipriano GP, Blaga LA, Dos Santos JF, Vilaça P, Amancio-Filho ST (2018) Fundamentals of force-controlled friction riveting: part II—joint global mechanical performance and energy efficiency. Mater 11(12):2489. https://doi.org/10.3390/ma11122489
16. Borba NZ, Afonso CRM, Blaga L, Dos Santos JF, Canto LB, Amancio-Filho ST (2017) On the process-related rivet microstructural evolution, material flow and mechanical properties of Ti-6Al-4V/GFRP friction-riveted joints. Mater 10(2):184. https://doi.org/10.3390/ma10020184
17. Borba NZ, Blaga L, dos Santos JF, Canto LB, de Traglia Amancio-Filho S (2018) Influence of rotational speed on the microstructure and mechanical performance of friction-riveted thermosetting composite joints. Weld Int 32(1):11–21. https://doi.org/10.1080/09507116.2017.1347321
18. de Proença BC, Blaga L, dos Santos JF, Canto LB, de Traglia Amancio Filho S (2017) Friction riveting ('FricRiveting') of 6056 T6 aluminium alloy and polyamide 6: influence of rotational speed on the formation of the anchoring zone and on mechanical performance. Weld Int 31(7):509–518. Retrieved from https://www.tandfonline.com/action/showCitFormats?doi=10.1080%2F09507116.2016.1218627

Part VI
Tooling and Process Monitoring

Linking Tool Features to Process Forces

Samuel Merritt, Ken Ross, and Yuri Hovanski

Abstract Tool design in linear friction stir welding has been largely Edisonian with organizations having favorite tools and features that have been used without clear data showing cause and effect. Previous studies evaluating features have elucidated certain trends with respect to reducing reactionary forces on the tools, although most of these studies focused on fixed tool diameters with few variations based on the complexity associated with machining or grinding new tool geometries. The present work focuses on FSW tools that have been produced using direct laser sintering to enable rapid evaluation of numerous tool features and designs. Tools were produced showing the effects of variation in shoulder convexity, shoulder features, probe features and probe geometry. Analysis of how each feature influences the transverse and axial forces are presented, as a means of demonstrating low-force tool designs.

Keywords Tools · Process forces · Friction stir welding

Introduction

Developed by TWI and patented in 1991, linear friction stir welding (FSW) operates by driving a rapidly rotating tool linearly through the joint line of the materials being welded. FSW is a solid-state process, meaning that the melting point of the material being joined is never reached. Rather than relying on fusion as a joining process, FSW facilitates diffusion between the materials being joined. Diffusion refers to the distribution of atoms between materials. Diffusion will occur naturally, though often very slowly unless ideal conditions are achieved. The rate of diffusion can be increased with high temperature, high pressure, and the availability of clean bonding surfaces. These conditions are created during FSW. Heat is generated by the

S. Merritt · Y. Hovanski (✉)
Brigham Young University, Provo, UT, USA
e-mail: yuri.hovanski@byu.edu

K. Ross
PNNL, Richland, WA, USA

Y. Hovanski et al. (eds.), *Friction Stir Welding and Processing XII*,
The Minerals, Metals & Materials Series,
https://doi.org/10.1007/978-3-031-22661-8_15

frictional force of the rotating tool rubbing against the material and energy released as the material is deformed [1]. Pressure is provided by the downward force of the FSW machine driving the tool into the welded material. Atomically clean surfaces are constantly provided as the shearing motion of the rotating tool mixes the materials being joined [2]. The design of FSW tools has a significant effect on the volume of material displaced and the motion of the material during the weld.

There are two major considerations when designing and selecting tooling for a friction stir weld: tool material and tool geometry. Tool material is critical because the material selected must be appropriate for the desired material to be welded. The melting temperature of the tool material must be well above the temperature that will be reached during the welding process. It is also preferable to select a hard material that will resist wear and deformation. Generally, steel tools are selected when working in aluminum or magnesium, and tungsten or PCBN tools are selected when working in steels or titanium [3].

The geometry of the tool plays an important role in the mixing of the welded materials. Poorly designed tooling can lead to defects in the weld such as tunneling or poor surface quality [4]. A current challenge in FSW is that often specific situations require specialized tooling. Creating appropriate tools for varying situations can be a costly, but necessary component of developing a new weld. The shoulder of the tool is responsible for the majority of material deformation, while the pin primarily assists with the vertical movement of material [5]. The features found on the shoulder and pin can vary widely based on purpose and preference.

The tool geometry that will be used in this study is commonly known as the CS4 style of tool. The CS4 consists of a concave, scrolled shoulder, and a conical probe with a step spiral. Previous research has found significant relationships between weld forces and tool features. One notable study found the following relationships to be the most significant [6].

- A larger probe cone angle was found to increase forge force.
- A larger shoulder radius was found to decrease both transverse and forge forces.
- A larger shoulder length increased the transverse and forge forces.

Previous studies focused on tools with a 25.4 mm diameter tool. Currently, this size of tool is the most commonly used in industry. Little research has been done to investigate whether the same relationships exist when smaller tools are used.

In this study, we will be developing tools following the CS4 tool geometry, though at a reduced size of 12.7 mm in diameter. The primary purpose of this study is twofold. First, to assess the effectiveness of using additively manufactured tools to understand the effects of tool geometry on weld forces. Second, to evaluate the relationship between tool parameters and weld forces for tools of this smaller size to see how they compare to previous research. The forces we will be analyzing are the downward force on the tool, or forge force, and the force acting in the direction of motion of the weld, or transverse force.

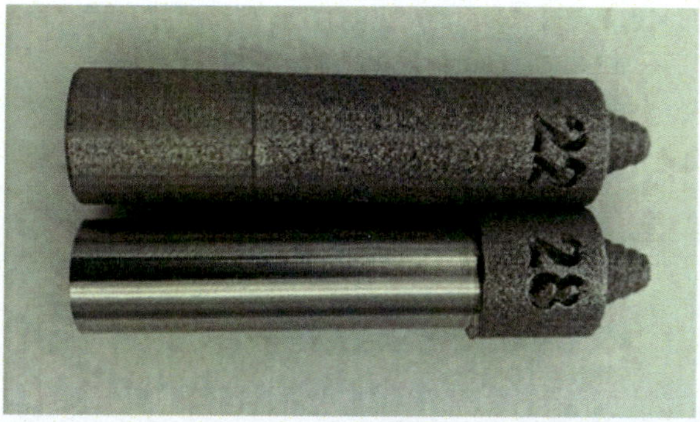

Fig. 1 Image depicting the two tools, tool 28 which has been turned down and tool 22 which has not

Methodology

Materials

All welds were completed in AA 7075-T6.

Tool Processing

The tools for this study were created from an austenitic stainless-steel powder (316L) via direct laser sintering. Using additive manufacturing processes allowed for full design freedom and greater complexity than other manufacturing methods. The tools were created using a Concept laser M2 Cusing. The tools were oriented on the build plate so that the shaft of the tool was perpendicular to the plate's surface, with the head of the tool facing upwards. The Concept laser operates by spreading a thin layer of metal powder and then using a laser to melt the new layer and create a solid where desired. As a result, the tools are fused to the baseplate upon completion. The finished tools are removed from the build plate using a bandsaw. Part of the shaft of each tool was then turned down from 12.7 to 10.66 mm using a CNC mill for the purpose of fitting the tools in a collet used during the welding process (Fig. 1).

Tool Design

Tool designs were developed using a parametric model in SolidWorks for ease and consistency. Figure 2 shows the parameters available for adjustment. Parameters not visible in the figure include the number of revolutions on the shoulder and the number of revolutions on the probe (Table 1).

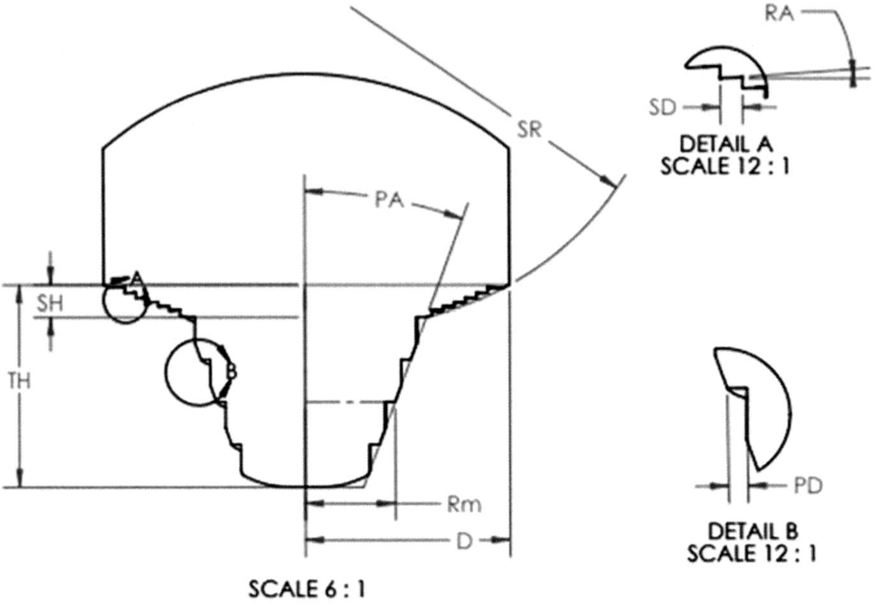

Fig. 2 Detail showing the parameters that can be adjusted in the parametric model

Table 1 Description of tool variables evaluated in this study

Parameter description	Abbreviation
Tool diameter	D
Tool height	TH
Shoulder radius	SR
Shoulder height	SH
Mid probe radius	Rm
Probe cone angle	PA
Revolutions on shoulder	O-rev
Revolutions on probe	I-rev
Shoulder spiral depth	SD
Probe spiral depth	PD
Shoulder spiral rake angle	RA

Welding Setup and Parameters

All welds were done on a TTI Linear RX friction stir weld machine using depth control. The welds were bead on plate with a length of 260 mm. A force limit of 8896 N was employed within the code to avoid damage to the tools or equipment. The welding parameters varied between the initial and the second test, as shown in

Table 2 Table showing linear FSW parameters used for evaluation of tools in this study

Weld parameters for initial tool set	Weld parameters for second tool set
• Plunge RPM: 800 • Linear RPM: 240 • Weld speed: 40 mm/min • Plunge depth: 6.15 mm • Tilt angle: 2°	• Plunge RPM: 1000 • Linear RPM: 400 • Weld speed: 40 mm/min • Plunge depth: 6.16 mm • Tilt angle: 2°

Table 2. These changes were made to decrease the amount of tool deformation seen during the weld and improve weld quality.

Results and Discussion

Accuracy of Printed Tools

The accuracy of the features present on the printed tools is a key aspect in validating the results of this study. We need to have confidence that changes to the parameters are correctly represented in the tool. If not accurately represented, the use of additive manufacturing to create prototype tools would be invalidated. Similarly, tool features need to be accurately created to determine what relationships exist between tool parameters and forces. A Keyence digital microscope was used to measure features such as shoulder height (SH) and depth of cut on the probe (PD). Comparing measured values to the expected measurements from the model revealed that the shoulder height and probe depth of cut were accurate to around $+/-$ 0.08 mm. Visual inspection confirmed the correct number of revolutions were present on both the probe and the shoulder. We were unable to gather measurements related to features on the shoulder such as rake angle due to the size of the tool and the rough surface created by the printing process.

Initial Tool Results

Our initial tool study consisted of ten tools featuring variations in probe cone angle, the number of revolutions on the shoulder and probe, and the rake angle as shown in Table 3. It was determined that each tool could only be used to complete one weld due to the amount of deformation experienced by the tool during the process. The average transverse and forge force during the linear portion of the weld were recorded and compared to the varied parameters for each tool. The results are displayed in Figs. 3, 4, 5, 6, 7, 8, 9, 10.

Table 3 Table showing the range of values to be studied for each parameter during the second trial

	Range of values
Shoulder height	0.5–1 mm
Revolutions on the probe	1.5–4 Rev
Revolutions on the shoulder	0.75–2.5 Rev
Rake angle	3–6 Degrees

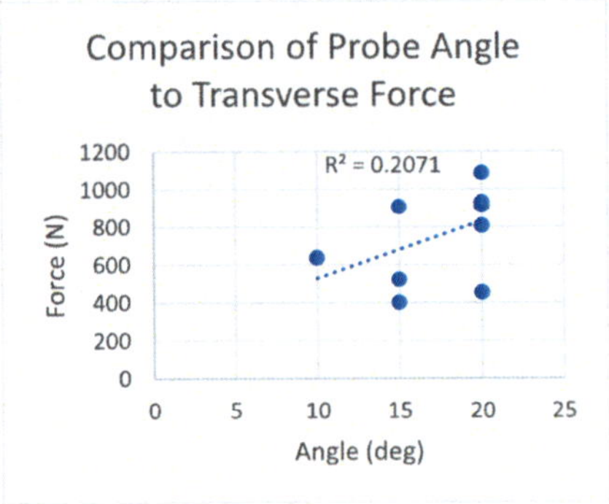

Fig. 3 Graph comparing the effect of probe cone angle on average transverse force

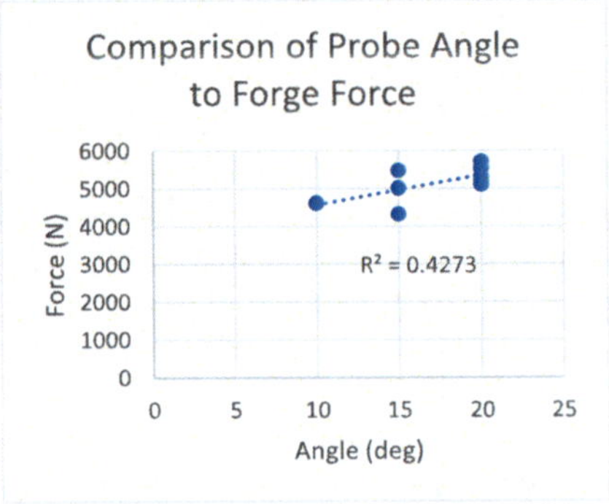

Fig. 4 Graph comparing the effect of probe cone angle on average forge force

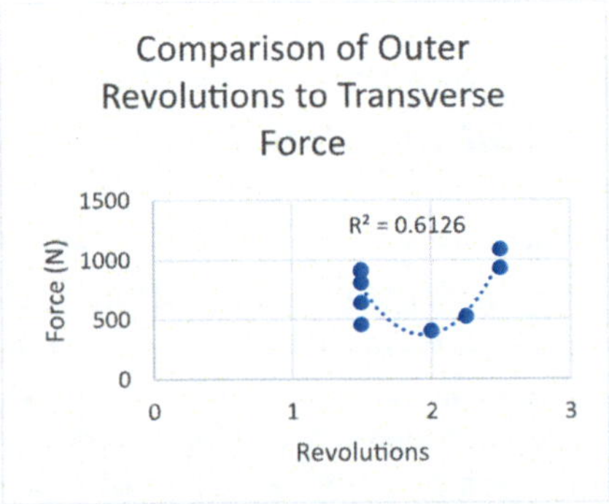

Fig. 5 Graph comparing the effect of the number of revolutions on the shoulder on average transverse force

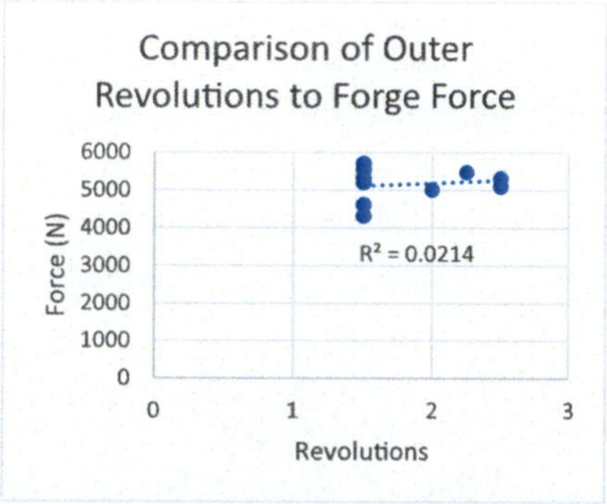

Fig. 6 Graph comparing the effect of the number of revolutions on the shoulder on average forge force

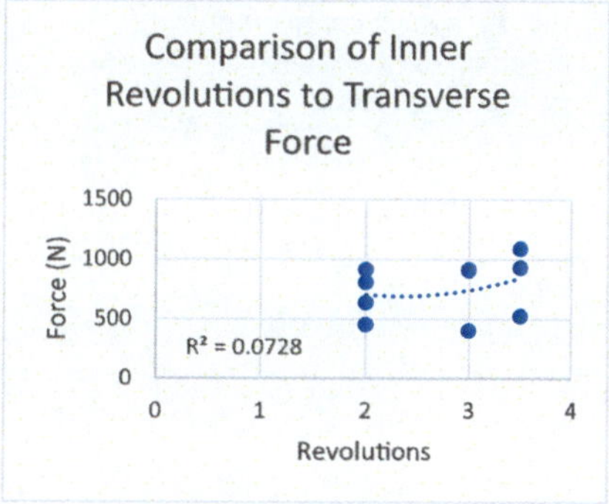

Fig. 7 Graph comparing the effect of the number of revolutions on the probe on average transverse force

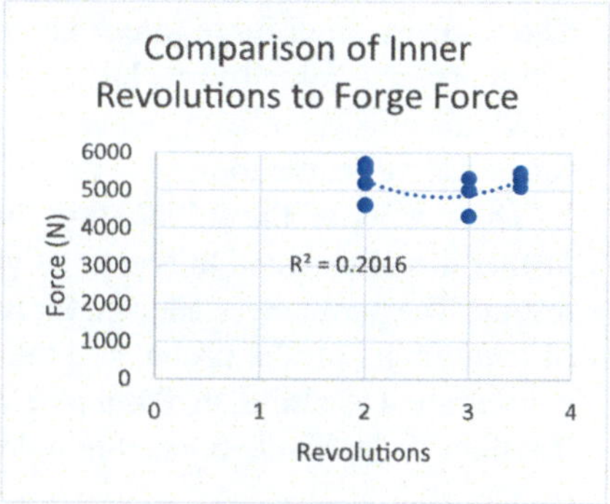

Fig. 8 Graph comparing the effect of the number of revolutions on the probe on average forge force

Looking at the trend line and R squared value for each of these plots, there are some potential correlations between the varied features and the weld forces. From our graphs, we can infer that there is a potential relationship between probe cone angle, the number of revolutions on the probe, and the rake angle, to forge force. There is also potentially a relationship between the probe cone angle and the number of revolutions on the shoulder to transverse force. Not all these relationships are linear.

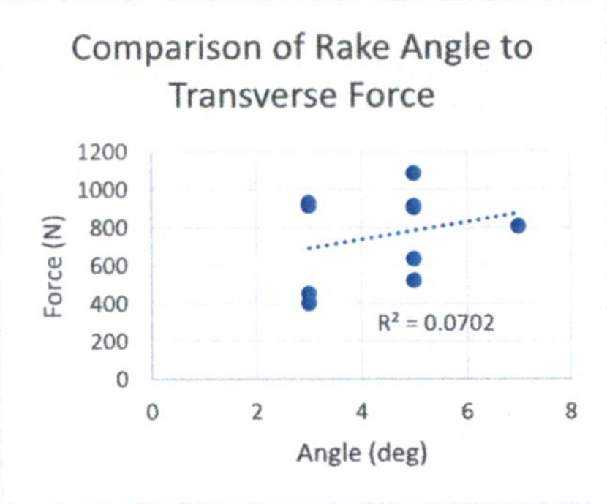

Fig. 9 Graph comparing the effect of rake angle on average transverse force

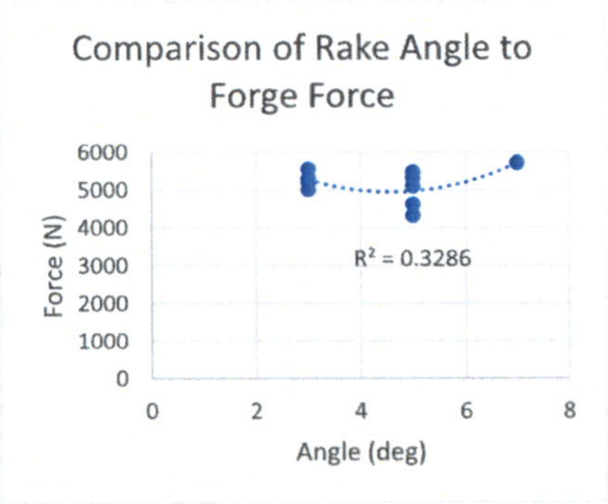

Fig. 10 Graph comparing the effect of rake angle on average forge force

The comparison of forge force to rake angle, forge force to revolutions on the probe, and transverse force to revolutions on the shoulder follow a parabolic relationship. This indicates there is an ideal value for these parameters that will minimize the forces acting on the tool.

The results of the initial study helped us to identify parameters we wished to further investigate and the range of values to use for these parameters in subsequent testing. The parameters selected for further investigation were rake angle, the number of revolutions on the probe, and the number of revolutions on the shoulder. Probe cone angle was eliminated as a parameter due to the tendency of tools with an angle less than 20 degrees to break. Shoulder height was selected as an additional parameter of interest.

Tool Results

The second set of tools developed consisted of 17 tools. Two copies were made of each tool so two sets of data could be recorded for each design. The tool designs were developed by selecting a master design from the previous study and varying one parameter for each subsequent tool. The average transverse and forge force for each weld was taken during the linear portion of the weld. Figures 11, 12, 13, 14, 15, 16, 17, 18 show the comparisons between weld forces and tool parameters.

Overall, we see stronger correlations and trends in our second round of testing. Creating multiple tools within a range of possible parametric values created a more comprehensive overview of the possible relationships. The strongest and clearest

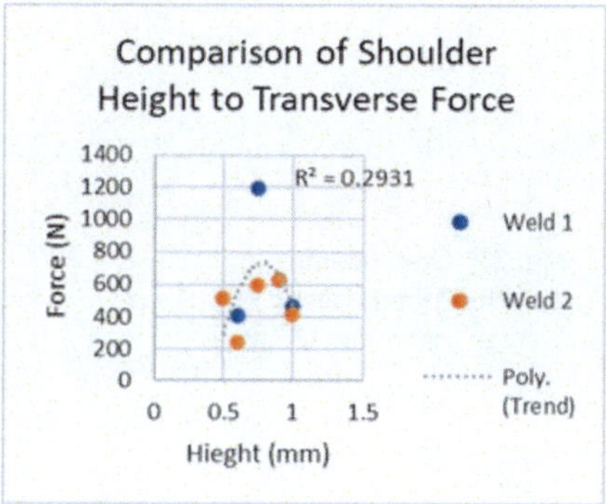

Fig. 11 Graph comparing the effect of shoulder height on average transverse force

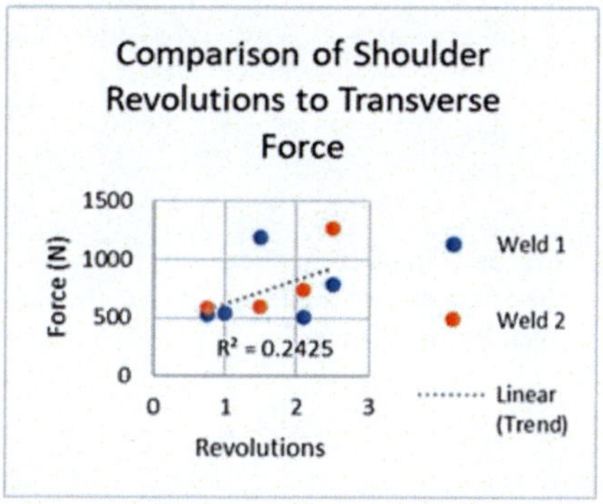

Fig. 12 Graph comparing the effect of the number of revolutions on the shoulder on average transverse force

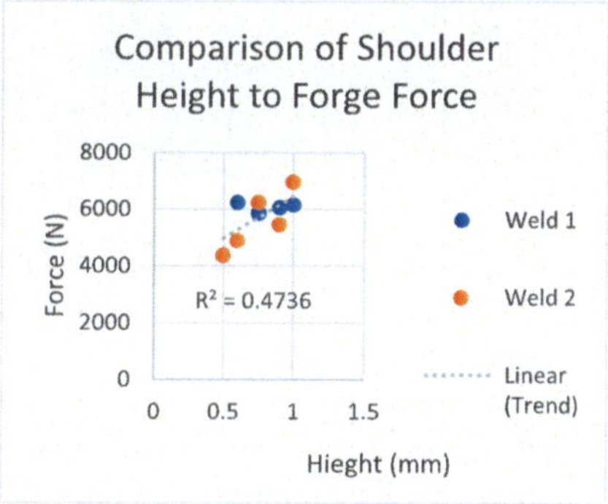

Fig. 13 Graph comparing the effect of shoulder height on average forge force

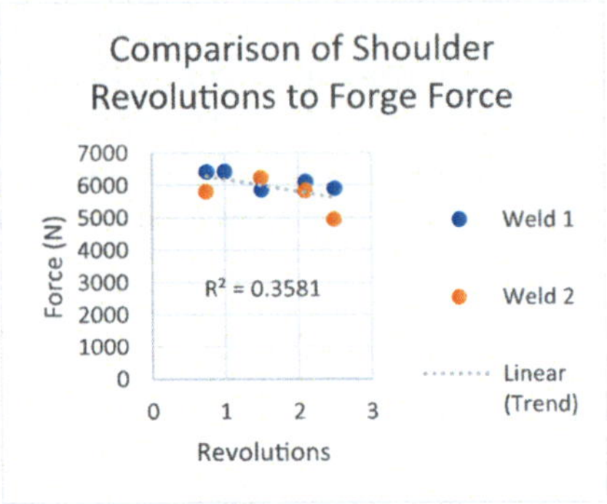

Fig. 14 Graph comparing the effect of the number of revolutions on the shoulder on average forge force

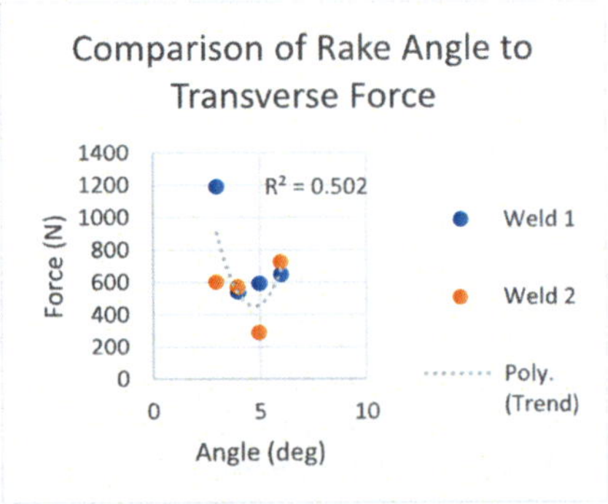

Fig. 15 Graph comparing the effect of rake angle on average transverse force

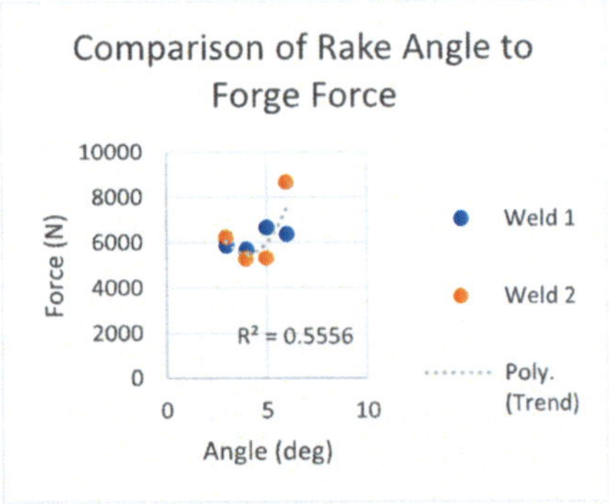

Fig. 16 Graph comparing the effect of rake angle on average forge force

correlations revealed are those relating shoulder height to forge force, the number of revolutions on the shoulder to forge force, and the number of revolutions on the probe to forge force.

The relationship between shoulder height and forge force indicates that a taller shoulder tends to increase the forge force during the weld. This seems plausible as the higher shoulder facilitates greater shoulder engagement during the weld, increasing

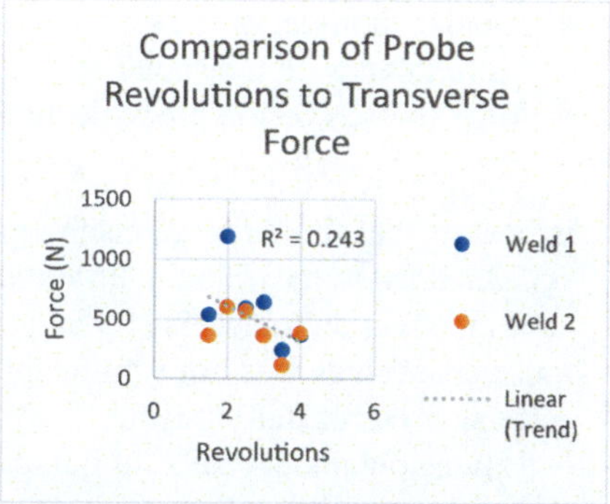

Fig. 17 Graph comparing the effect of the number of revolutions on the probe on average transverse force

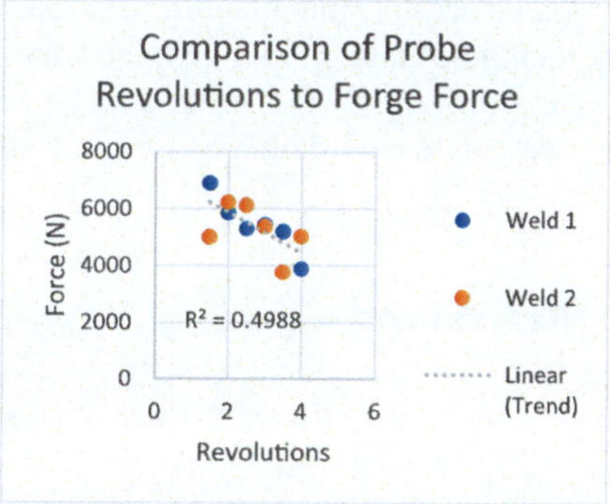

Fig. 18 Graph comparing the effect of the number of revolutions on the shoulder on average forge force

the amount of contact and material displacement experienced by the tool. As such, an increase in forge force should be expected.

The results of this study would also indicate that increasing the number of revolutions on the tool shoulder and the probe tends to decrease the forge force. This likely related to the increasingly fine thread we see as we increase the number of revolutions. Courser threads are more aggressive in material deformation, requiring greater amounts of force.

Other parameters may have effects on weld forces, but it is difficult to draw conclusions due to noise in the data. It is possible that some of the finer features such as rake angle were lost when working with tools of this scale and using an additive process.

Conclusions

From the results of this study, the following conclusions can be drawn.

- A larger probe cone angle increases forge force.

- A taller shoulder increases forge force.
- More threads on the shoulder decrease forge force.
- More threads on the probe decrease forge force.

The first two conclusions echo the results seen in the previous CS4 tool study, indicating that similar relationships exist between tool features and weld forces in the smaller sized tools. There is little indication that the tool features used in this study have an effect on transverse force during the weld. It is possible that the variation in transverse forces was not significant enough to be detected by the data acquisition systems used for this study.

Having obtained results similar to other studies validates the use of additive manufacturing in tool development and testing. Research in this area can be costly and difficult due to the expense and time required to acquire tools with unique features. Using additive manufacturing to rapidly create tools offers a cheaper and more timely method of testing tool geometry and its affects. It does appear that complexity can be a limiting factor when dealing with tools of a smaller size, tool details such as rake angle would likely be more visible on a larger tool.

References

1. Schmidt HB, Hattel JH (2008) Thermal modelling of friction stir welding. Scripta Mater 58(5):332–337
2. Kumar K, Kailas SV (2008) The role of friction stir welding tool on material flow and weld formation. Mater Sci Eng, A 485(1–2):367–374
3. Rai R, De A, Bhadeshia HKDH, DebRoy T (2013) Review: friction stir welding tools. Sci Technol Weld Joining 16(4):325–342
4. Trueba L, Heredia G, Rybicki D, Johannes LB (2015) Effect of tool shoulder features on defects and tensile properties of friction stir welded aluminum 6061–T6. J Mater Process Technol 219:271–277
5. Kadian AK, Biswas P (2017) Effect of tool pin profile on the material flow characteristics of AA6061. J Manuf Process 26:382–392
6. Nielsen BK (2009) Developing response surfaces based on tool geometry for a convex scrolled shoulder step spiral (CS4) friction stir processing tool used to weld AL 7075

Friction Stir Welding Operating Window for Aluminum Alloy Obtained by Temperature Measurement

Moura Abboud, Laurent Dubourg, Adrien Leygue, Guillaume Racineux, and Olivier Kerbrat

Abstract Finding the process parameters (advancing and rotational speed) in FSW is the key to construct an operating window based on a type of material and a tool geometry. Literature shows different process methods carried out by researchers and engineers to obtain the data needed to select suitable parameters. These methods can be divided into "destructive", based on macrostructures analyses, and "non-destructive" with temperature measurement. This article describes a continuous temperature measurement method of the nugget zone using a thermocouple embedded in the tool for aluminum alloy. The results show that the FSW window obtained by temperature measurement is wider than the one obtained by analyzing the macrostructures. It is then possible to quickly acquire the technological window of operating parameters adapted to the tool-material couple. The aim of this study is to reduce time of the welding process, number of trials, cost and guarantee a good quality.

Keywords Aluminum · Modeling and simulation · Process technology

M. Abboud (✉) · O. Kerbrat
UMR 6183, Université de Rennes, ENS Rennes, CNRS, 35000 Rennes, GeM, France
e-mail: moura.abboud@ens-rennes.fr

O. Kerbrat
e-mail: olivier.kerbrat@ens-rennes.fr

M. Abboud · L. Dubourg
Stirweld, 35170 Bruz, France
e-mail: laurent.dubourg@stirweld.com

A. Leygue · G. Racineux
UMR 6183, Nantes Université, Ecole Centrale de Nantes, CNRS, 44321 Nantes, GeM, France
e-mail: adrien.leygue@ec-nantes.fr

G. Racineux
e-mail: guillaume.racineux@ec-nantes.fr

Y. Hovanski et al. (eds.), *Friction Stir Welding and Processing XII*,
The Minerals, Metals & Materials Series,
https://doi.org/10.1007/978-3-031-22661-8_16

Introduction

Friction stir welding (FSW) is a solid-state joining process invented by the Welding Institute (TWI) in 1991 [1]. It has now reached an important level in industrial sector where it is applied in mass production due to its environmental advantages and energy efficiency. This technique consists of assembling two parts in three steps using a non-consumable rotating tool. First, plunging of the tool into the workpiece, then dwelling during which the tool rotates to generate heat for a specific time duration to form a pasty state, and finally welding where the tool is given a certain advancing velocity to travel along the joining line to form the joint by stirring the material (Fig. 1) [2].

Heat generated during FSW process results from two phenomena: (1) plastic deformation of the material around the tool, depending on advancing and rotational speed, and (2) friction between the tool and the material governed mainly by the FSW welding tool geometry (Fig. 2), forge force (F_z Fig. 1), and rotational speed.

However, weld quality control is a challenging task due to temperature variation throughout the weld. In particular, the temperature has an important influence on the porosity rate which could dramatically decrease mechanical properties of the welded part.

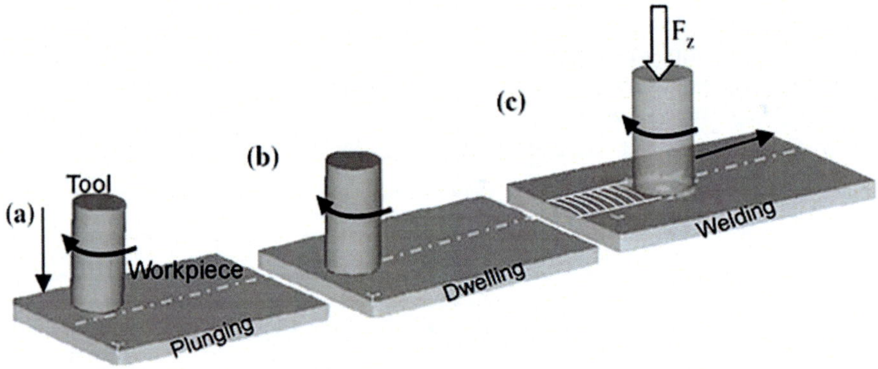

Fig. 1 Different steps of FSW process [2]

Fig. 2 Different geometry of FSW tools

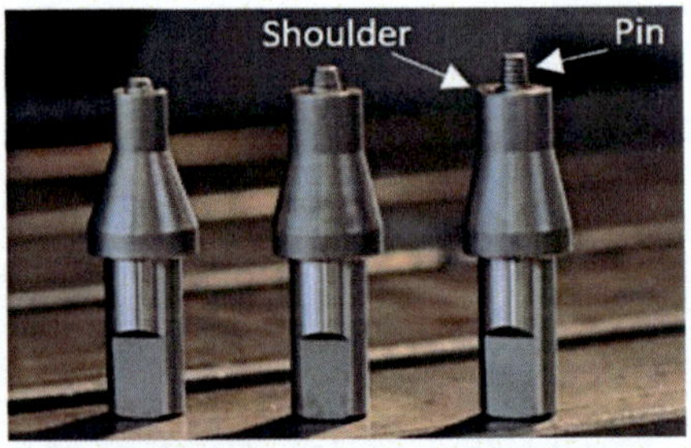

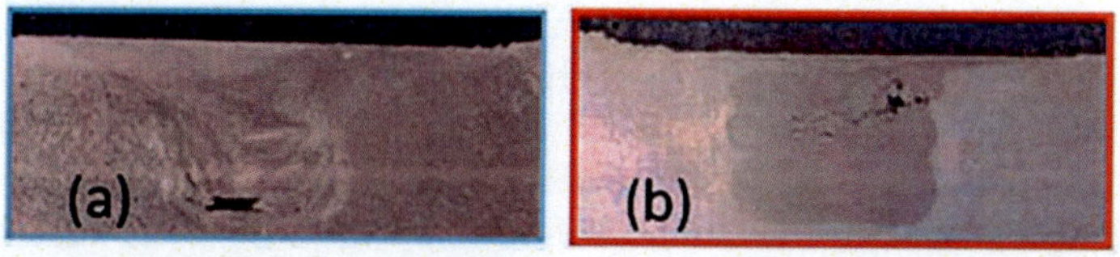

Fig. 3 Porosity defects: **a** cold weld, **b** hot weld

According to the ISO-25239 standard, porosity defects of a weld caused by the evolution of the temperature can be divided into two parts:

- Cold weld whose porosities are at the bottom of the welding (Fig. 3)
- Hot weld whose porosities are at the top of the weld (Fig. 3).

The temperature increases if the speed of rotation increases and decreases when the speed augmented lead [3]. So FSW parameter setting represents a critical point to obtain a decent quality weld.

The objective of the FSW process is to increase the performance of assemblies, the productivity, and guarantee optimum quality, considering all the parameters that affect the temperature which in turn determines the quality of a weld.

For this reason, researchers and engineers show interest in finding the good parameters (rotational and advancing speed) of a welded piece without defects. Several experimental methods could be used to obtain the welding process window.

In Table 1, various methods to characterize a welded joint for several types of aluminum alloy are described.

The difficulty is to have a good temperature measurement which allows to qualify a weld. In using the innovative technique of temperature measurement, operating windows are obtained and delimited to show the different weld areas. Thus, Sect. 2 presents the state of art of temperature measurement, Sect. 3 is dedicated to the tests conducted, and Sect. 4 shows the results obtained on two grades aluminum alloys (AA-6082 and AA-5754) before a discussion (Sect. 5) and a conclusion (Sect. 6).

State of Art

Operating Window Obtained by Temperature Measurement

Thermocouples embedded into the workpiece make it possible to measure the temperature in each area of the FSW seals depending on their fixing position [11]. Spindle rotation and plastic deformation of the material can affect the position of the thermocouple in the part. Silva et al. [12] and De Backer et al. [13] proposed an intense study with several positions of the thermocouple.

The use of thermocouples integrated in the tool has been studied and applied with success in the FSW process. The thermocouple is usually embedded through a hole drilled in the tool and positioned as close as possible to the tool-part interface providing faster response to disturbances during welding. The direct contact of thermocouple with the base metal was reported by [14] who used two thermocouples integrated in the tool.

The thermal camera above the parts to be welded [15] makes it possible to perform several tests without having to modify the parts. However, there are some limitations

Table 1 Different methods used to qualify a weld

Experimental methods	Description	References	Pros/Cons
Macrostructural analysis	The welded workpiece is cut in samples using a cutting tool, then machined to the required dimensions for the macrostructure specimens. After that, the specimens are polished using different grades of emery papers. Macrostructural analysis are carried out using a light optical microscope incorporated with an image analyzing software. The microstructure of the experimental sample is compared with reference microstructures	[4] [5] [6]	• Macrostructural results are 100% sure and follow the same rule of defect • Applicable method for all types of welding (butt and lap) • Many numbers of tests to characterize the defect-free region • Cost in time of preparation and analysis of macrostructural specimen • Piece cannot be used after the test; therefore, it is called destructive method
X-ray scanning	X-ray scanning is used to visualize deviations and anomalies in the structure and surface of a material, as well as up to 30 cm behind the surface. The images are generated in real time. X-ray pixel width is set to a certain number, so flaws smaller than this size is not detectable by the scan	[7] [8]	• The ability to control the pixel width to detect the flaws • Using this procedure, signal-to-noise ratio increases which means better specifications since there is more useful information than unwanted data • Non-destructive method of the piece • Applicable for butt joint • The cost of an additional machine to detect the defects

(continued)

Table 1 (continued)

Experimental methods	Description	References	Pros/Cons
Temperature measurement	Several temperature measurement techniques have been explored by researchers to measure or predict the temperature of the weld, by using: 1. Thermocouples embedded in the workpiece: Mounted perpendicularly and rather close to the surface where the temperature needs to be measured 2. Thermocouples embedded in the FSW tool: Inserted through a drilled hole in the tool and positioned as close as possible to the interface tool-workpiece 3. Thermal infrared camera: Fixed orthogonally to the plate. The information obtained by thermal camera is for the surface Based on literature, the peak temperature of the material should not exceed its solidus temperature. The temperature increases with the increase of rotational speed and decrease of advancing speed It is possible to weld in good conditions, with the same tool, in an area between $0.7T_f$ and $0.8T_f$	[9] [10] [11]	• Knowing that an operating window can be delimited to three regions (hot weld, sound weld, and cold weld), boundaries can be set to separate these areas based on a function of solidus temperature • Calculating the value of the limit temperature separating the cold-sound area and the hot-sound area allow the tests to be carried on until the limited temperature is reached • Based on temperature measurement method, time and cost of tests can be reduced by eliminating the macrostructural analysis and x-ray scanning • In case of small diameter tools, the number of thermocouples is limited and difficult to be located in the desired positions • In case of IR camera, quality problems are due to the emissivity of aluminum and the difficulty of fixing the camera orthogonally to the part • Destructive method when it comes to thermocouple embedded in the workpiece

related to the measurement areas and problems associated with the quality of the measurement.

In the literature, there is a lack of information about the relation between the temperature and the defects. Gratecap et al. used thermocouple embedded in the tool to find that the window of operating parameters of Al 6000 series extends between $0.7T_f$ and $0.8T_f$ [9]. Okamura et al. showed for the same Al series that the weldability zone is limited between $0.75T_f$ and $0.8T_f$ [16]. D. Ambrosio used all the three methods on AA5083-H111, AA-6082-T6 and AA7075-T6 to find their operating window for a rotational speed range 500 rpm and 1750 rpm and advancing speed equal to 120, 240, and 360 mm/min. Based on the temperature measurement, a sound weld for the three alloys is obtained for a temperature higher than 472 ± 5 °C [7]. Imam et al. reported a defect of cold weld when temperature in AA6063-T4 is lower than 350 °C [17]. Fehrenbacher et al. showed internal defects of AA6061-T6 when the temperature is lower than 515 °C [14].

"Stirweld" Innovative Temperature Measurement System

In this context, an innovative temperature measurement system integrated into the tool has been developed for the Stirweld company (that develops FSW welding heads) in order to measure the welding temperature in real time using a system of transmission by cable, and the position of the thermocouple allows its integration into tools of different diameters while remaining in a position as close as possible to the welding.

Experimental Details

A HAAS-VF3 CNC machine with Stirweld head of 3500 rpm and a critical force of 25 kN was used to conduct cord welding of the aluminum lloy. Tools were machined from the steel-based alloy X37CrMoV5-1 H11. The tool dimensions used in this study are: shoulder diameter 8.5 mm with a conical pin of 4 mm diameter at the base, 3 mm at the top and a length of 3 mm (Fig. 4 Tool dimensions).

This tool is drilled to be able to insert the thermocouple inside the tool to measure the temperature. The distance between the thermocouple and the weld is optimized due to a precise position of the thermocouple in the tool. The thermocouple used in this experimental campaign is of type J of 2 mm diameter.

The base metals used are AA-6082 T6, a structural alloy with an average strength but excellent corrosion resistance commonly used for machining, having a solidus temperature of 555 °C, and AA-5754 H11, an alloy with a high resistance having a solidus temperature of 610 °C. The alloys are in the form of rectangular sheets with dimensions $595 \times 120 \times 4$ mm.

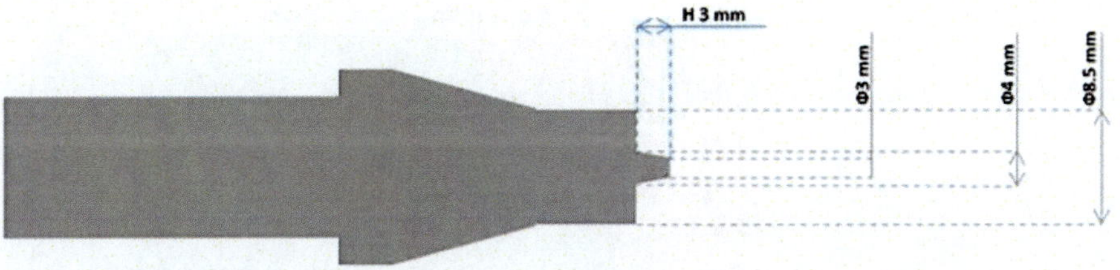

Fig. 4 Tool dimensions

The first pair of welding parameters (advancing speed; rotational speed) is chosen by finding a good contact between the shoulder and the material to find the corresponding axial force: for a hot weld (with flash), the axial force is decreased, and for a cold weld (not enough contact between the tool and the workpiece), the force is increased with respect to the maximum force of the head. After having the parameters for a good visual welding, a 250 mm long weld is made by varying the speed of rotation ω between 980 and 3500 rpm and the advancing speed V between 100 and 1000 mm/min for the AA-6082 (Table 2) and the rotation speed ω between 980 and 3500 rpm and the advancing speed V between 25 and 2250 mm/min for AA-5754 (Table 3).

Based on the literature (Fig. 5), the material's temperature should not exceed its solidus temperature (between $0.7T_f$ and $0.8T_f$ for a defect-free weld) and it increases with the increase of rotational speed and the decrease of advancing speed. Considering this statement and to choose the next pair of parameters, the rotational speed is maintained, and the advancing speed is increased (1). Then start the welding process and observe the temperature, if it is too low, the axial force is increased until a good

Table 2 Welding parameters for AA-6082

V (mm/min)	100	200	294	378	420	504	630	1000
ω (rpm)	2100 2800 3500	980 1260 1420 1680 2100 2800 3500	980 1260 1420 1680 2100 2800	980 1260 1420 1680 2100 2800	980 1260 1420 1680 2100 2800	980 1260 1420 1680 2100 2800	980 1260 1420 1680 2100 2800	1260 1420 2100 2800

Table 3 Welding parameters for AA-5754

V (mm/min)	25	50	100	420	630	1000	1500	1750	2000	2250
ω (rpm)	3500	980 1420 2800 3500	980 1420 2800 3500	980 1420 2800 3500	980 1420 2800 3500	2800 3500	2800 3500	2800	2800 3500	2800 3500

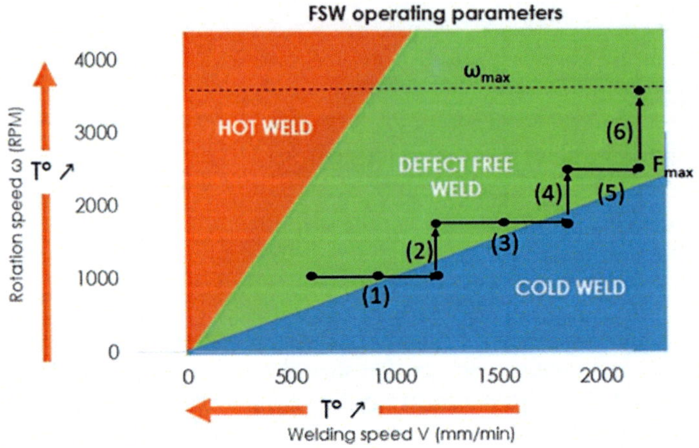

Fig. 5 Experimental approach to choose FSW operating parameters

visual welding is obtained, and after that, the advancing speed is increased another time. The tests are carried on in a way that when the force limit is reached, the advancing speed is maintained, and the rotational speed is increased (2). The cycle is repeated until the maximum rotational speed (6) is reached and the maximum force is reached (5).

To be sure that the weld is safe without defects and to find a relation between the parameters and the solidus temperature, a polishing of the section at 200 mm is needed (according to ISO-25239). First, the piece is sliced using an electrical saw. Then, the specimens are coated using thermosetting resin to facilitate the handling of the parts and to preserve intricate edges and surface defects during metallographic preparation. Macrostructural analysis was accomplished using a USB digital microscope (RS-PRO) incorporated with an image analyzing software (Micro-Capture Plus) The specimens were polished using different grit size of emery papers and final one is the diamond compound (3 μm particle size). For a better macrostructure vision, the sections were attacked with sodium hydroxide to reveal the trace of the tool.

Results

Metallographic Examination

The metallographic views of the two aluminum grades are shown in Figs. 6 and 7.

Welds with porosity at the bottom represent cold welds (blue outline), welds with porosity at the top represent the hot welds (red outline), and welds without porosity represent a good quality (green outline). We thus distinguish an operating window

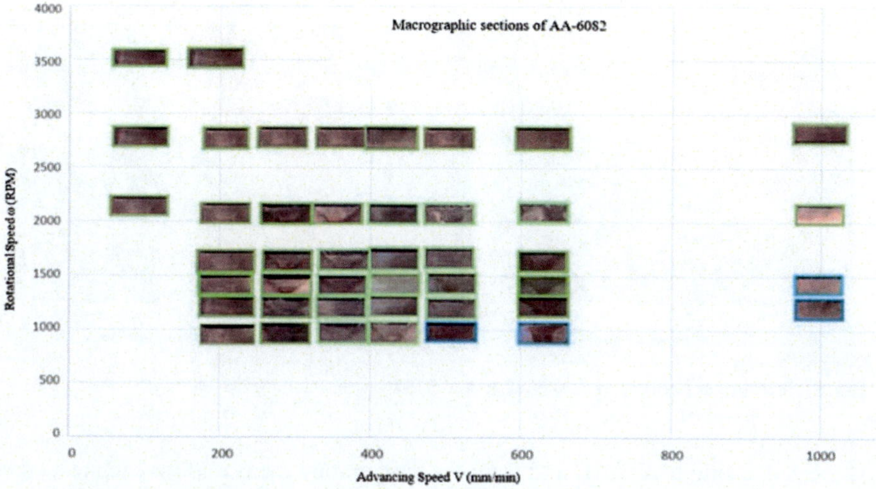

Fig. 6 Macrographic sections of AA-6082

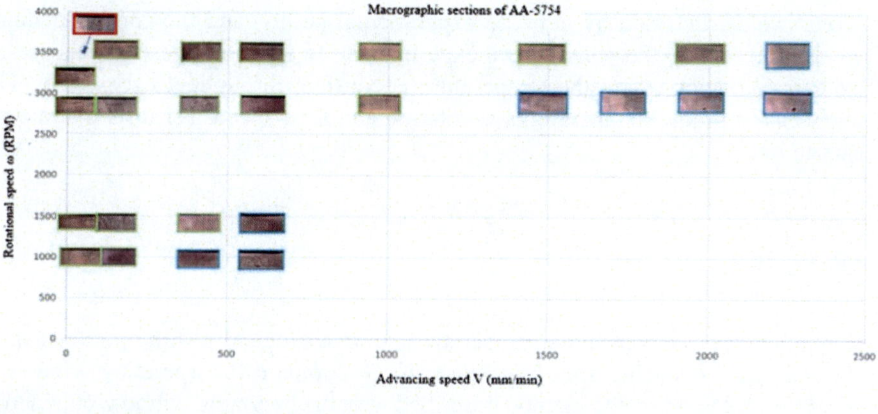

Fig. 7 Macrographic sections of AA-5754

for each alloy, allowing subsequently to optimize the manufacturing parameters in order to improve productivity (speed) while guaranteeing a "sound" weld.

Temperature Curves of AA-6082 and AA-5754

The temperature variation is continuously measured using the thermocouple embedded in the tool. The temperature used is that where the macrographic section

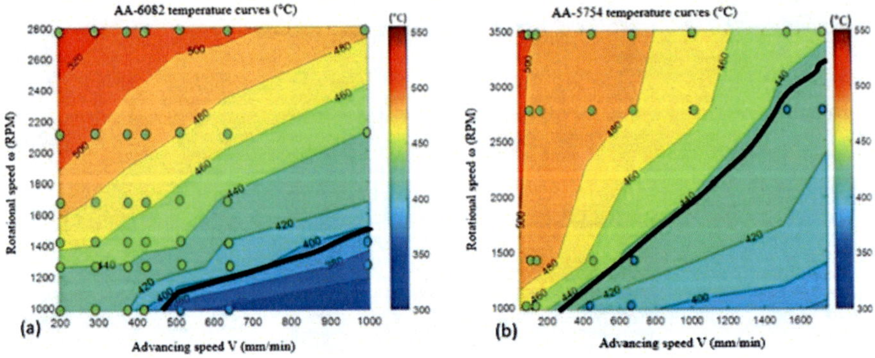

Fig. 8 Temperature curves: **a** AA-6082 **b** AA-5754

is located. Using MATLAB and from the temperature measured for each test (speed in advance; rotational speed), the isothermal curves are plotted and shown (Fig. 8).

These curves show that the maximum temperature does not reach never the solidus temperature of the material [18]. Limit (black curve) exists between the zone of the "cold" welds (detected by the blue experimental points) and the zone of "sound" welds (detected by the points green experiments). In addition, these graphs present isothermal curves where the temperature is constant throughout each curve. This allows us to predict temperatures at rotation and feed speeds for tests that are not carried out.

Discussion

For both types of material, the temperature increases logically with rotational speed increase and advancing speed decrease. If we compare the operating window of AA-6082 obtained by the method described with the operating window obtained by [7], the temperature of a sound weld should be higher than 390 °C as for the second method it was supposed to be higher than 477 °C which makes the operating window for AA-6082 wider for the innovating solution. We can also compare AA-5754-H111 and AA5083-H111; the temperature of a sound weld should be higher than 440 °C as for the other method it was supposed to be higher than 477 °C.

The innovating solution has an advantage over the methods presented in the article. It is easy to use without the need to be a professional. The difference in the cost between a conventional tool (traditional method) and the tool used in the temperature measurement is less than 20%. As for the methods with temperature measurements, it does not need a hole the tool holder which makes it easier to change the tool without the tool holder (reduce cost) and makes it compatible with any thermocouple diameter.

Finding the boundaries between sound weld and the defective welds, it is easier to rely on the temperature measurement method to gain time and reduce the number of trials.

Conclusion

In this article, we experimented with a technique that allows us to measure the real-time weld temperature and qualify a weld quickly, so to obtain operating windows. By using this technique, one can reduce the cost of the tests by reducing their number and adapting this technique as a replacement for the polishing and cutting process macrographic. In fact, carrying out a small number of tests with temperature measurement makes it possible to visualize the operating window and the limit between cold welds and sound welds. Also, the cost of the temperature system can be reduced by the fact that we do not need to drill a hole in the tool holder, and in the tool, also we do not need more than one thermocouple placed in the right position to measure the temperature. The choice of parameters can then be made by seeking to increase productivity (increase advancing speed) without going to the side of the cold welds. Thus, tests at very high speeds of rotation and feed can be carried out remaining within the operating parameters. It is even applicable on very complex parts such as cold plates with a fine geometry.

Nevertheless, a certain number of parameters, such as the geometry of the tool, the force forging, and the thickness of the material are not considered in our study. These aspects will be studied in the rest of our work to be able to generalize the use of this method.

Acknowledgements This work was supported by the start-up company "Stirweld" as a collaboration on a research project.

References

1. Gb SG et al (1994) Improvements relating to friction welding. In: European Patent Office EP0653265A2
2. Chauhan P, Jain R, Pal SK, Singh SB (2018) Modeling of defects in friction stir welding using coupled Eulerian and Lagrangian method. J Manuf Process 34(June):158–166. https://doi.org/10.1016/j.jmapro.2018.05.022
3. Adriana M, Obregon T (2011) Effect of process parameters on temperature distribution, microstructure, and mechanical properties of self- reacting friction stir welded aluminum alloy 6061-T651. In: NMC Inf Serv Concur Technol Corp, Harrisburg, PA, USA
4. Seeman M, Sivaraj P (2019) Comparison of friction stir welding window for AA 2024 and AA 6082 alumunium alloys. Int J Manag Technol Eng IX:4017–4027. 16.10089.IJMTE.2019.V9I01.18.27873

5. Gadakh VS, Kumar A (2014) Friction stir welding window for AA6061-T6 aluminium alloy. Proc Inst Mech Eng Part B J Eng Manuf 228(9):1172–1181. https://doi.org/10.1177/0954405413510289
6. Seetharaman R (2015) Friction stir welding process window for AA 7075 aluminium alloy. IJCRT 3(1):314–321. ISSN 2320–2882
7. Ambrosio D (2022) A semi-empirical model for peak temperature estimation in friction stir welding of aluminium alloys. Sci Technol Weld Join. https://doi.org/10.1080/13621718.2022.2065106
8. Saravanan T (2014) Non-destructive evaluation of friction stir welded joints by X-ray radiography and infrared thermography. Procedia Eng 86:469–475. https://doi.org/10.1016/j.proeng.2014.11.060
9. Gratecap F, Racineux G, Marya S (2008) A simple methodology to define conical tool geometry and welding parameters in friction stir welding. Int J Mater Form 1(3):143–158. https://doi.org/10.1007/s12289-008-0370-z
10. Colegrove PA, Shercliff HR (2003) Experimental and numerical analysis of aluminium alloy 7075–T7351 friction stir welds. Sci Technol Weld Join 8(5):360–368. https://doi.org/10.1179/136217103225005534
11. Yuqing M, Liming K, Fencheng L, Yuhua C, Li X (2016) Investigations on temperature distribution, microstructure evolution, and property variations along thickness in friction stir welded joints for thick AA7075-T6 plates. Int J Adv Manuf Technol 86(1):141–154. https://doi.org/10.1007/s00170-015-8182-z
12. Silva ACF, De Backer J, Bolmsjö G (2017) Temperature measurements during friction stir welding. Int J Adv Manuf Technol 88(9–12):2899–2908. https://doi.org/10.1007/s00170-016-9007-4
13. De Backer J, Bolmsjö G, Christiansson AK (2014) Temperature control of robotic friction stir welding using the thermoelectric effect. Int J Adv Manuf Technol 70(1–4):375–383. https://doi.org/10.1007/s00170-013-5279-0
14. Fehrenbacher A, Duffie NA, Ferrier NJ, Pfefferkorn FE, Zinn MR (2014) Effects of tool-workpiece interface temperature on weld quality and quality improvements through temperature control in friction stir welding. Int J Adv Manuf Technol 71(1–4):165–179. https://doi.org/10.1007/s00170-013-5364-4
15. Lambiase F, Paoletti A, Di Ilio A (2018) Forces and temperature variation during friction stir welding of aluminum alloy AA6082-T6. Int J Adv Manuf Technol 99(1–4):337–346. https://doi.org/10.1007/s00170-018-2524-6
16. Okamura H (2000) Visualization of material flow in an autogenous friction stir weld. J Jpn Weld Soc 69(7):565–571
17. Imam M, Biswas K, Racherla V (2013) On use of weld zone temperatures for online monitoring of weld quality in friction stir welding of naturally aged aluminium alloys. Mater Des 52:730–739. https://doi.org/10.1016/j.matdes.2013.06.014
18. Chao YJ, Qi X, Tang W (2003) Heat transfer in friction stir welding—experimental and numerical studies. J Manuf Sci Eng 125(1):138–145. https://doi.org/10.1115/1.1537741

The Performance of a Force-Based General Defect Detection Method Outside of Calibration

Johnathon B. Hunt and Yuri Hovanski

Abstract Friction stir welding (FSW) is a solid state joining process that is suitable for many engineering designs. The need for economical non-destructive examination (NDE) for welding is imperative for high volume industries. This need has led the development of many different inline force-based NDE methods. This work introduces the performance of a generalized NDE method that is applied to welds that were not included in the calibration. When this NDE method is applied to welds within calibration, the process spatially detects defects with 94% accuracy with 4% false positives. In addition, the methodology proved 100% effective at positive detection when defects were present with zero scrap rate as a Go, No Go test.

Keywords Friction stir welding · Non-destructive evaluation · Non-destructive testing · Forced based

Introduction

Many current engineering designs use the friction stir welding (FSW) process to join metals. FSW is particularly beneficial for hard to weld metal alloys such as certain grades of aluminum because FSW avoids solidification and liquidation cracking because the welding temperature is below the melting temperature. FSW as a "green" welding process uses less power than traditional arc welding methods and does not use shielding gases or other consumables such as filler wire [1]. Automotive OEMs have taken advantage of these FSW benefits to add more aluminum in their products. To achieve the high production volumes in the automotive industry, many have researched and developed welding practices to achieve high-speed friction stir welding [2]. With high-speed welding technologies in place, there are thousands of friction stir welds apart of today's vehicles. As more vehicle manufacturers adopt friction stir welded joints, their suppliers will need to increase outputs of their supply

J. B. Hunt (✉) · Y. Hovanski
Brigham Young University, Provo, UT, USA

Y. Hovanski et al. (eds.), *Friction Stir Welding and Processing XII*,
The Minerals, Metals & Materials Series,
https://doi.org/10.1007/978-3-031-22661-8_17

chains. Reducing the cycle time of a friction stir welded joint could provide additional throughput suppliers need to meet demands. One possibility to reduce cycle time is by integrating non-destructive examination (NDE) into the welding process. This small time savings could have a large impact allowing more aluminum parts to be introduced into future vehicles [3]. Currently, manufacturers of friction stir welded joints use ultra-sound or x-ray to validate weld quality. These processes are traditionally a post weld examination, increasing the cost to produce a friction stir welded joint.

Force-based NDE methods have been researched to facilitate inline FSW inspection and reduce these costs [4–8]. These methods use the measured welding forces to validate weld quality. Most of the methods in these studies are limited in their application within the respective training parameters. Limiting the ability to be applied in different welding environments. This limitation requires training for any changes to the machine, FSW tool, or any welding parameter.

A more generally applied NDE method has been previously presented that has shown the ability to validate weld quality while allowing changes in welding speed as well as tool position [9]. This method was applied in six different ways. First, the Y axis or Z axis forces were independently used to locate defects along the length of the weld, as defined in Fig. 1. Next, the Y and Z axes forces were combined to locate defects along the weld. Then the Y axis and Z axis forces were independently used to validate if the weld had any defects, without providing the location along the weld. Finally, the Y and Z axes forces were combined to determine if a weld had any defects. The performances of these six applications of this NDE method can be found in Table 1. To further understand the capability of this general forced-based NDE method, this work will quantify the performance when welds were produced in a welding environment that was outside of the calibrated environment.

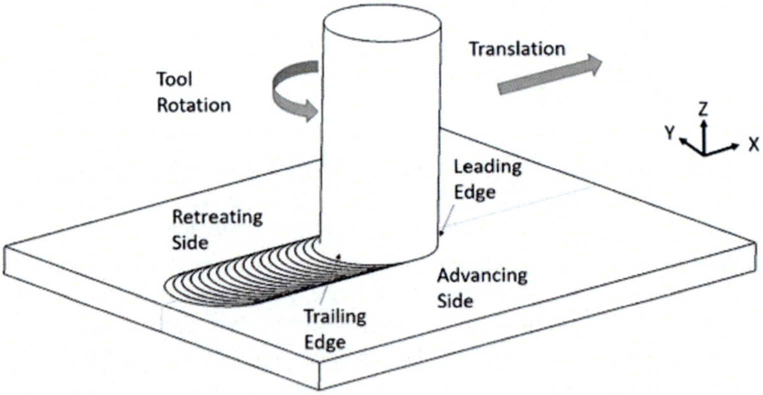

Fig. 1 Shows the work coordinate system referred to throughout this document

Table 1 Includes the performance of the six applications of the general force-based NDE method. This performance is quantified by using the standard confusion matrix terminology of true positive rate (TPR), or the rate the method correctly predicts a defect, avoiding type II error, and false positive rate (FPR), or how frequently the method results in a type I error

Metric	Y axis with location (%)	Z axis with location (%)	Y & Z axis with location (%)	Y axis without location (%)	Z axis without location (%)	Y & Z axis without location (%)
TPR	70	80	94	78	94	100
FPR	0	11	4.2	22	11	0

Methodology

Thirty 3.8 mm thick AA5754 welded samples built the sample set for this work. Figure 2 depicts an example weld of the 495 mm long welds. The different welding parameters used can be found in Table 2 and all variations shared a constant spindle speed of 1600 RPM (26.67 Hz). A specifically designed FSW machine, supported by Bond Technologies, made all the welds. The workpieces were constrained with manual clamping and is depicted in Fig. 3. The FSW tool included a flat 12 mm in diameter with two scroll started shoulder. The tool pin was threaded with three flats 120 degrees apart. The base diameter of the pin was 6 mm and was 3.1 mm long including a 10-degree taper. The tool material was heat treated H13. All welds were produced in a bead on plate configuration with the tool pin smaller than the material thickness, to protect the FSW tool and machine. The Cartesian direction forces and spindle torque were sampled at a frequency of 1250 Hz by the native DAQ measuring system. Forces were measured by three Kistler 9078C tri-directional load cells that are located in between the machine gantry and the spindle head as shown in Fig. 4. During a typical friction stir weld, there are four different force segments. First, the tool plunges of into the material where there is an initial spike of forge force followed by a decrease once the material has softened. As the machine ramps up to the commanded welding speed down the weld seam, the tool passes into cooler material where forging forces increase after plunge. Then as the tool comes out of the ramp, the forge force reaches a pseudo steady state part of the weld, and that force is maintained through the duration of the weld. Finally, the tool exits the material and forge force is relieved from the tool. These four segments are displayed in Fig. 5. The NDE method only uses the steady state portion of the force data. Thus, the data was truncated and included the data after to the ramp to just before the exit. Next, the data was binned into smaller lengths by the MATLAB spectrogram function which calculates the power spectral density (PSD) of the force of the respective measured signal [10]. Smaller force bins facilitate force arrays have a nonvariant mean. The smaller the bin size also creates finer time steps between PSD calculations which provides more test points along the length of a weld. PSD calculations, denoted as $X_{\mathrm{PSD}}(f)$, are equivalent to the squared FFT amplitude, $|X(f)|$ and then normalized by the size of the frequency bin, N_f. This relationship can be observed in Eq. (1). In tandem to the PSD calculations, 2D radiographic images, from Avonix Imaging

MN, validated that the sample set included volumetric defects. These x-ray images provided the spatial location of any volumetric defects. An example of the 2D x-rays is found in Fig. 6. Equally spaced distance defect logic vectors were built accordingly, with "1" interpreted as defect and "0" as defect-free for each weld in the sample set. After radiographic imaging, cross-sectional metallography quantified the area of a defect.

$$X_{\mathrm{PSD}}(f) = \frac{|X(f)|^2}{N_f} \tag{1}$$

An Olympus SZX12 running Leco Paxit 2 software optical microscope captured the images of the cross sections. Defect sizes ranged between 0.003 to 0.6 mm^2. However, the x-ray images were only capable of defining voids with a width of approximately 0.27 mm^2 or larger. Using optical micrographs at the extremes of the x-ray images defect definition, defect heights ranged from 0.08 to 0.6 mm. Thus, the x-ray images were able to define volumetric voids whose areas were approximately 0.02 mm^2 or larger.

To understand the performance of the PSD threshold NDE method outside of the calibrated welding parameters, four different combinations of the sample sets produced thresholds to test against the rest of the sample set. These different combinations are grouped by welding speed. The group denoted by "A" calibrated the threshold and the "B" group used the calibrated threshold to predict defect presence. Details of the four groups are found in Table 3. The calibrated thresholds were built for each directions with 100 evenly spaced thresholds between the minimum and

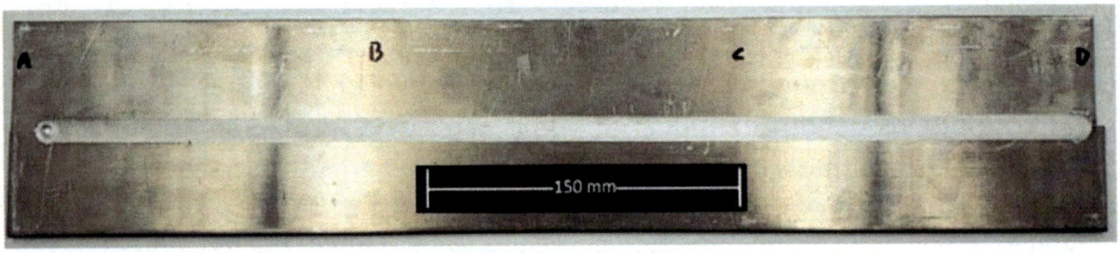

Fig. 2 Example of a friction stir weld. This specific weld was produced with parameter set 4 in Table 2

Table 2 Traverse velocity and welding tilt varied throughout the experiments and all combinations are included here

Set Number	Transverse Speed (mm/min)	Tilt (deg)	Qty
1	1500	2	5
2	2000	2	
3	1500	1	
4	2000	1	
5	2500	1	
6	3000	1	

Fig. 3 Clamping method and welding table set up for all welds

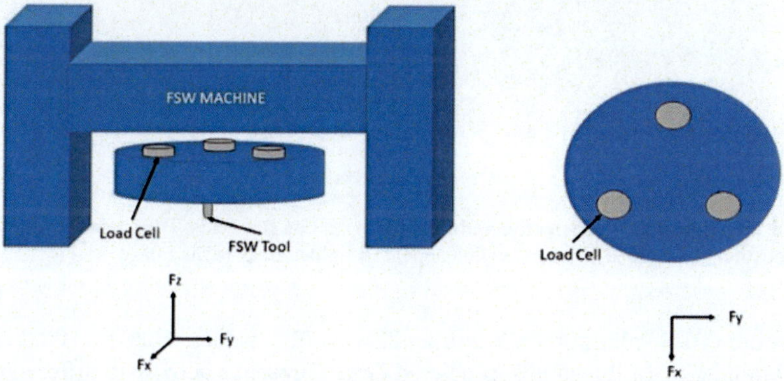

Fig. 4 Force measurement system in the RM FSW machine with three triaxial load cells located between the FSW tool and the FSW machine cross beam

Fig. 5 Green area represents the data truncation used in this study. The plot includes only the Z axis force from a weld that was produced with parameter set 1 in Table 1

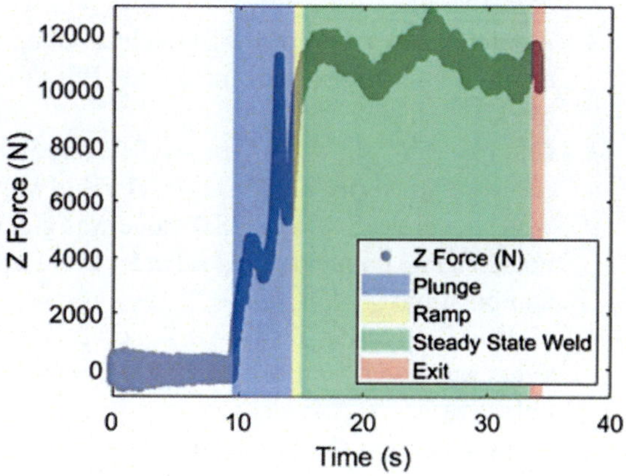

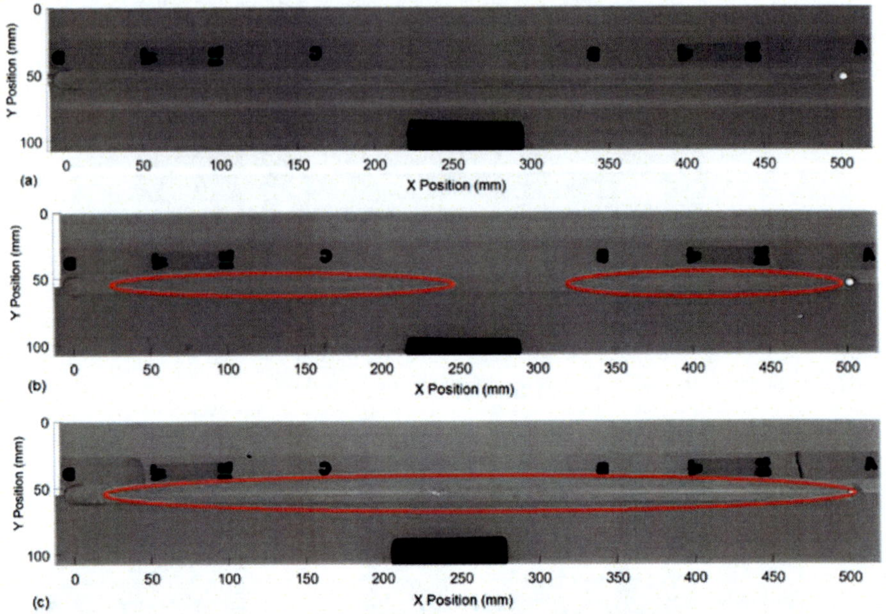

Fig. 6 Example x-rays of defective and defect-free friction stir welds. The circled regions highlight the location of volumetric defects which appear as lighter areas in the image

maximum PSD values of the Y axis and Z axis PSD values. Then all 100 thresholds, or combination of thresholds, predicted defect presence across six different tests:

1. Y Axis Test – All Y axis PSD values along a weld were compared to verify if a value was less than the Y axis PSD threshold. If not, then that point was marked defective.
2. Z Axis Test – All Z axis PSD values along a weld were compared to verify if a value was greater than the Z axis PSD threshold. If not, then that point was marked defective.
3. Combined Y & Z Axes Test – All Y and Z axes PSD values along a weld were compared to verify if a Y axis PSD value was less than the Y axis PSD threshold **AND** to verify if a Z axis PSD value was greater than the Z axis PSD threshold. If not, then that point was marked defective. Repeated for all Y and Z axes threshold combinations.
4. Y Axis Go, No Go Test – All Y axis PSD values were compared to verify if all value was less than the Y axis PSD threshold. If not, then that weld was marked defective.
5. Z Axis Go, No Go Test – All Z axis PSD values were compared to verify if all value was greater than the Y axis PSD threshold. If not, then that weld was marked defective.
6. Combined Y & Z Axes Go, No Go Test- All Y and Z axes PSD values were compared to verify if all the Y axis PSD values were less than the Y axis PSD

Table 3 Details of the four sets of data that built threshold that were applied outside of the parameter sets

Group name	A group parameter sets	Welding speeds mm/min	B group parameter sets	Welding speeds mm/min
AB	1–4	1500 & 2000	5 & 6	2500 & 3000
ABA	1, 3, 6	1500 & 3000	2, 4, 5	2000 & 2500
BA	5 & 6	2500 & 3000	1–4	1500 & 2000
BAB	2, 4, 5	2000 & 2500	1, 3, 6	1500 & 3000

threshold **AND** to verify if all of the Z axis PSD values were greater than the Z axis PSD threshold. If not, then that weld was marked defective. Repeated for all Y and Z axes threshold combinations.

The resultant predictions were then compared to the 2D x-ray images to build confusion matrices for each threshold. TPR and FPR were calculated from the confusion matrices. Each TPR and FPR combination was then compared to the perfect performance of 100% TPR and 0% FPR to find the minimum distance between the perfect performance and actual performance. The threshold that had the minimum distance between performances was used to predict defect presence for group B respectively for each test.

Results and Discussion

The six different tests across the four different groups resulted in 24 resultant TPR and FPR when predicting defect presence outside of the calibrated welding parameters. Figure 7 displays the results for each of the four groups. As shown in Table 1, the best performing tests when the test sample welding parameters was included in the threshold calibration were combined Y & Z axes PSD values, and Go, No Go tests for Z axis and combined Y & Z axes PSD values. However, only the AB group followed this trend. In addition, Group AB had the most consistent performance where three of the six different tests resulted in 100% TPR and 0% FPR. The other three groups had a much larger spread. The Z axis PSD test was consistent through all of the different groups with the average performance of 88% TPR and 8.9% FPR. However, these results are drastically different than the performance of the respective performance within the calibration.

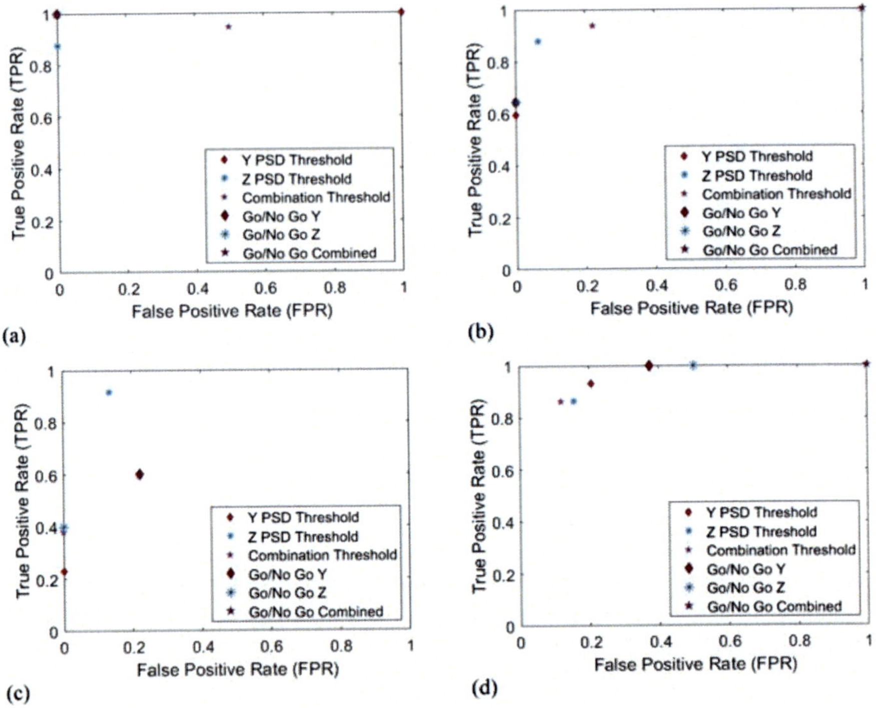

Fig. 7 Performance of the best threshold from group A applied to group B. **a** contains the results for the AB combination. **b** includes the results from the ABA combination, **c** contains the results for group BA, and **d** contains the results for group BAB

Conclusion

Calibrated Y axis, Z axis and combination of Y and Z axes PSD thresholds predicted defect presence in welds whose welding parameters were not included in the calibration. Four different calibrations and predictions sets were calculated and compared. If can be concluded that:

- General defect detection Go, No Go tests can be applied outside of the calibrated areas with 100% confidence when the calibration included welding speeds within or slower than the welds in question.
- The application of the Z axis PSD threshold was the most robust when applied outside of its calibration with the worst performance of 86% TPR and 16% FPR and best performance of 87% and 0% FPR.
- In general, the worst application of calibrated PSD thresholds is when welds in question are slower than the welds used for the calibration.

References

1. Akinlabi ET, Akinlabi SA (2012) Friction stir welding process: a green technology. In: Proceedings of world academy of science, engineering and technology, no 71. World Academy of Science, Engineering and Technology (WASET), p 1536
2. Hovanski Y et al (2015) High-speed friction-stir welding to enable aluminum tailor-welded blanks. JOM 67(5):1045–1053. https://doi.org/10.1007/s11837-015-1384-x
3. Hunt JB, Hovanski Y, Mazzeo B, Larsen B, Luther I (2021) Introduction to friction stir welding spectral defect detection, issue 5A-1. In: Sheet metal welding conference XIX
4. Boldsaikhan E, Corwin EM, Logar AM, Arbegast WJ (2011) The use of neural network and discrete Fourier transform for real-time evaluation of friction stir welding. Appl Soft Comput 11(8):4839–4846. https://doi.org/10.1016/j.asoc.2011.06.017
5. Das B, Bag S, Pal S (2016) Defect detection in friction stir welding process through characterization of signals by fractal dimension. Manufact Lett 7:6–10. https://doi.org/10.1016/j.mfglet.2015.11.006
6. Rabe P, Schiebahn A, Reisgen U (2022) Deep learning approaches for force feedback based void defect detection in friction stir welding J Adv Joining Process 5:100087. https://doi.org/10.1016/j.jajp.2021.100087
7. Franke DJ, Zinn MR, Rudraraju S, Pfefferkorn FE (2021) Real-time measurement of friction stir tool motion during defect interaction in aluminum 6061-T6
8. Mishra D et al (2020) Real time monitoring and control of friction stir welding process using multiple sensors. CIRP J Manuf Sci Technol. https://doi.org/10.1016/j.cirpj.2020.03.004
9. Hunt JB, Mazzeo BA, Sorensen CD, Hovanski Y (2022) A generalized method for in-process defect detection in friction stir welding. J Manufact Mater Process 6(4):80. https://doi.org/10.3390/jmmp6040080
10. Signal Processing Toolbox (2019) The Mathworks Inc., Natick, Massachusetts, United States. Accessed on 6 Sep 2022

The Role of Fracture Properties on Lap Joint Strength of Friction Stir Welded AA7055-T6 Sheets

Check for updates

Kranthi Balusu, Hrishikesh Das, Xiao Li, Timothy Roosendaal, Robert Seffens, Ayoub Soulami, and Piyush Upadhyay

Abstract Friction stir lap welded (FSLW) joints have weight-saving potential in aluminum-intensive automotive assembly. However, friction stir welding (FSW) also modifies the material microstructure close to the joint. Optimizing the FSLW joint strength requires understanding the relationship between the strength and the joint's local properties or microstructure. In previous studies, efforts have been dedicated to determining the effects of local softening, the shape of the oxide line, and porosity. However, the role of changes in fracture properties on the joint's strength has not been studied. In this work, fracture test procedures to characterize the fracture properties within the weld were proposed. The data from these fracture tests has been utilized to calibrate the parameters of the Gurson-tvergard-needleman (GTN) damage model within finite element analysis simulations. Using the weld fracture data, the simulation of a 3-sheet (aluminum alloys 7055–7055-6022) FSLW joint successfully predicted the lap shear strength to be within 10% of the experimental value. Comparison with strength prediction using only the base metal properties indicates that fracture property in the nugget region is crucial in determining the strength of AA7055 FSLW joints.

Keywords Aluminum · Characterization · Computational materials science and engineering · Fracture

Introduction

A key priority for the automotive industry is improving energy efficiency, and an important way to achieve that is through lightweighting in automotive structures. With such priorities, the adoption of high-strength 7xxx and 6xxx series aluminum alloys in structural components continues to grow [1, 2]. However, welding by traditional fusion processes is challenging for such aluminum alloys because they are susceptible to hot cracking and porosity [3]. In addition, for multi-material joints,

K. Balusu (✉) · H. Das · X. Li · T. Roosendaal · R. Seffens · A. Soulami · P. Upadhyay
Pacific Northwest National Laboratory, Richland, WA 99354, USA
e-mail: kranthi.balusu@pnnl.gov

Y. Hovanski et al. (eds.), *Friction Stir Welding and Processing XII*,
The Minerals, Metals & Materials Series,
https://doi.org/10.1007/978-3-031-22661-8_18

using conventional methods leads to the formation of intermetallic compounds and results in poor joint strengths [4]. In comparison, friction stir welding (FSW) can enable the joining of these challenging aluminum alloys, and it has been successfully used for various multi-material joints [4]. Additionally, the energy consumption of FSW is lower than fusion, thereby improving the joining process's energy efficiency [5].

Optimization of FSW process parameters and obtaining the strongest possible joint is not trivial. It is because the process-microstructure-property relationships in FSW are not comprehensively understood. During FSW, the material close to the weld undergoes large deformation at high rates and temperatures, resulting in grain recrystallization with hard-to-predict grain and dislocation structures [6]. Defects in the weld are also usually a concern [7]. Additionally, in age-hardenable alloys, such as 6XXX and 7xxx series aluminum, precipitate dissolution, reprecipitation, and over-aging occur [3]. The complex interplay of these mechanisms makes it an expensive task to understand even a few aspects of the process-material microstructure relationships [8]. Additionally, it is difficult to determine the microstructure-local/global property relationships within the weld because these microstructures are unique and they also vary within various regions of the weld [6].

An approach commonly taken to simplify this problem is to identify the aspects of local material properties or microstructures that are the most influential on the strength and focus process optimization on these aspects. For instance, it is now well known that in butt welds, the degree of softening due to the heat generated from the friction stir process is the most influential in determining its strength [9]. The effect of some defects, such as voids, on the strength is also known [10]. However, such studies still need to be carried out for other types of joints and aluminum material systems used. The aspects that control a friction stir lap welded (FSLW) joint's strength are less understood. The aspects affecting the FSLW strength can be expected to differ for each of the three different fracture modes encountered [11], making it a more complex problem. Shear fracture at the interface, tensile fracture originating at one of the ends of the interface between the two sheets, and tensile fracture in the HAZ are the three modes of fracture. The fracture mode could depend on the FSW parameters or, in general, the material system used. Typically, for similar lap joints in the 7xxx series, tensile fracture originating at the interface is the observed fracture mode [12]. For this fracture mode, the interface shape is known to be one of the aspects of the FSLW joint that affects its strength [11, 13], but studies into other aspects of the weld's local material microstructure/property are lacking.

To the best possible knowledge, the role of the fracture properties of a friction stir weld in determining the FSLW joint's strength remains uninvestigated. Fracture properties can be expected to be influential based on the knowledge from similar joints, such as the lap shear adhesive specimens [14] and friction stir spot welds [15]. However, fracture properties in FSW are not easy to characterize for multiple reasons. Firstly, the presence of large gradients and anisotropy in properties within the small weld means fracture sample sizes in the order of a few mm are needed. Designing fracture tests with such requirements would be expensive. An additional difficulty arises from the fact that fracture properties, unlike yield stresses, are not related to easily characterizable values such as hardness. Empirical relationships with hardness are available only in a few cases [16–18], and none exist for AA7055. More fundamentally, at the microstructural scale, the fracture toughness results from a complex interplay of multiple poorly understood mechanisms [19]. Therefore, microscopy can be ruled out as an option to determine fracture properties. The attempts made to measure fracture properties in friction stir welds either was for 6xxx-series aluminum [20–22] or AA7449 [17]. In one study, the determined fracture properties were used to predict joint strength, but it was for 6xxx-series aluminum butt joints [23]. However, no study has used the weld's fracture properties to predict the strength of a FSLW joint made from any material system.

In this paper, we first present a test procedure to determine fracture properties in the weld. Then fracture data is used to calibrate parameters for a damage model in the finite element analysis (FEA) simulations. In the next section of the study, we perform FEA simulations of the FSLW joint to predict its lap shear strength and compare it with experimental data.

Determination of Fracture Properties

Fracture Test Setup

A 3-sheet lap joint, as shown in Fig. 1, was manufactured using FSW. The shoulder diameter of the FSW tool used was 12.7 mm, and its grooves can be seen on the top surface in Fig. 1a. The tool was rotated at 1700 rotations per minute and moved in a straight line at a speed of 0.5 m/min. The weld was performed in the length direction of the 380 mm long sheets through almost the entire length while leaving a 12 mm gap with the edges. The 3-sheet configuration comprises two 2.5 mm thick sheets made of aluminum alloy 7055-T6 and one 1 mm thick Aluminum alloy 6022-T6 sheet stacked on top of each other. Figure 1b shows the cross-section of the 3-sheet stack-up. The FSLW joint is also visible in this figure. The darker region indicates the stir zone or the nugget zone, within which the three sheets were bonded. The figure also shows the dimensions of the joined areas between the sheets.

Single-edge notch bending (SENB) specimens were made from the weld for fracture testing based on ASTM standards E2818 and E1820. Figure 2a shows the

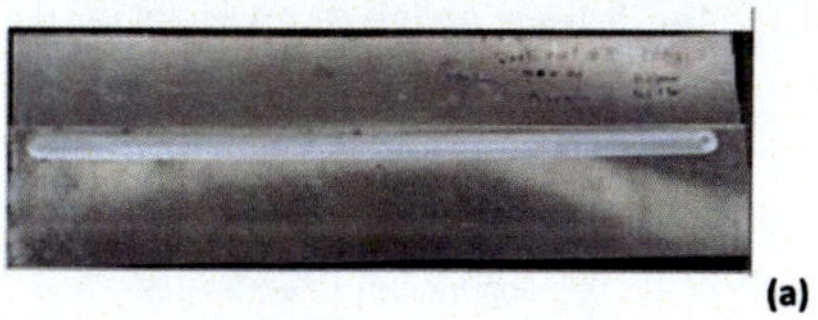

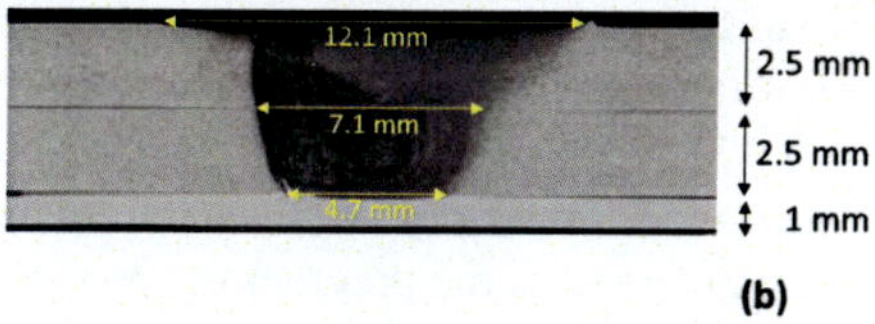

Fig. 1 3-sheet friction stir lap welded (FSLW) joint is shown using **a** the top view and **b** the view of the cross-section normal to the weld direction

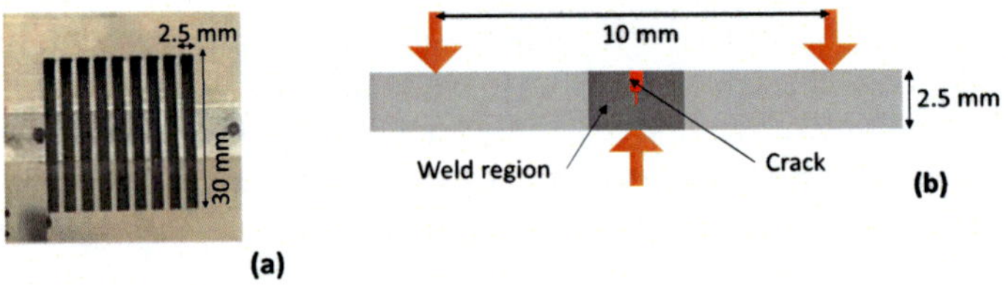

Fig. 2 Specimens used for single edge notch bending (SENB) fracture test; **a** lap joint after the specimens were cut out; **b** schematic of the specimen in the direction normal to the weld

top view of the lap joint after the fracture test specimens were machined using water jet cutting. The specimens were cut from locations centered on the weld line and away from the plunge and exit locations. The specimen's thickness (in the weld direction) is 2.5 mm, and the length is 30 mm. Additional machining involved removing the bottom two sheets, using a cut through the joint between the top two 7055 sheets. As shown in Fig. 2b, only the 2.5 mm thick top sheet remains. A fracture test specimen requires an already existing crack. Therefore at the center of this specimen, on the top edge, a notch was made using an EDM wire of radius 0.25 mm. The three notch lengths used in this study are 0.35, 0.45, and 0.5 times the width of the specimen. In addition to the specimen from the welded lap joint, fracture specimens were also cut from a pristine AA7055-T6 sheet to assess the fracture properties of the base metal. For this base metal specimen, the notch length was 0.35 times the width of the specimen. The significant change from the standards was the sample width, which was smaller because of the constraints of the joint design. As a result, this SENB specimen design neither confirms the plane strain nor plane stress conditions and therefore requires a 3D analysis.

This specimen was then mounted onto a 3-point bending machine, as shown in Fig. 3. The machine consists of three supports using which the sample can be loaded. The two stationary supports are on the edge of the specimen with the notch. These two supports were separated by a distance of four times the width of the sample, i.e., 10 mm is the span. The specimen was adjusted such that the notch was present at the center of this span. On the other edge, the third support also lies at the center of the span. This third support moves in a displacement-controlled way to bend the specimen.

The notch in the sample is not a good approximation of a crack because it has a finite radius. Therefore, an infinitesimally sharp crack was generated using fatigue precracking near the notch. This fatigue precracking process involved using a 165 N load for the shorter notch and a 100 N load for the longer notch at an R ratio of 0.1, as suggested in the standards. This procedure was carried out for around 10,000–20,000 cycles until a distinct drop in the cyclical load was visible. The samples were sectioned at the center of the sample thickness so that precracks were visible and the effectiveness of the precracking procedure could be verified. As Fig. 4 shows, sharp cracks originated from the notch and grew in the notch direction. The precracks were approximately 0.1–0.2 mm in length, with an average of 0.145 mm.

Fig. 3 Picture of the experimental test setup used to conduct the SENB test

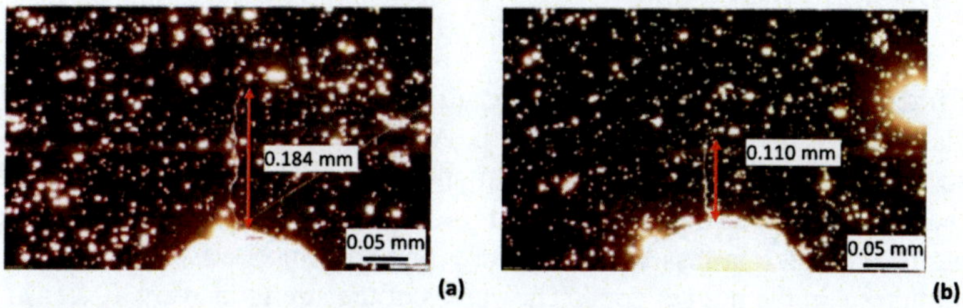

Fig. 4 Micrographs of the fracture test specimen cross-sections showing the presence of fatigue precracks extending from the notch; **a** longest and **b** shortest

All the fracture specimens were prepared in this described procedure. Then, the same test setup carried out fracture tests under monotonic loading. The specimens were gradually bent under quasistatic conditions until the specimens failed.

Simulation Setup for Calibration

The same fracture test was also simulated so that the experimental data could help calibrate the parameters of the damage model. The simulation was carried out using Abaqus Explicit FEA software. The 3D model of the SENB fracture specimen is shown in Fig. 5. The FEA model consists of a mesh with 0.1 mm-sided cube elements in the region close to the crack. The rest of the mesh was much larger, with a 0.5 mm dimension in the length direction. The cracks were modeled simply as separated/disjointed elements in the region corresponding to the crack plane. The supports were modeled as rigid bodies. Gurson-tvergard-needleman (GTN) damage

Fig. 5 The simulation model of the sample used for the SENB test

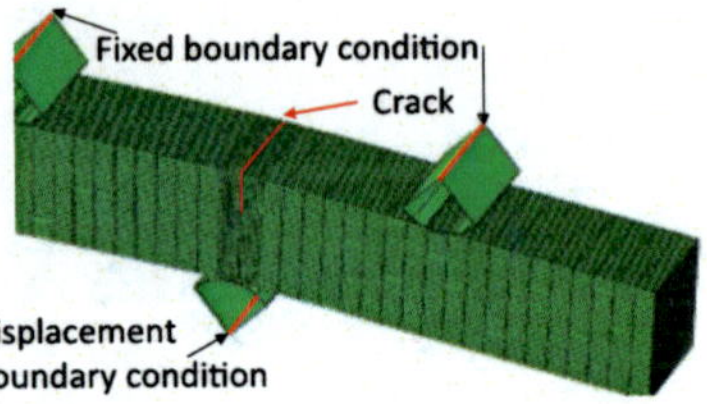

model, built in Abaqus software as porous metal plasticity and described in the appendix, was used to model the fracture. The elastic and plastic properties of 7055-T6 and 6022-T6 were sourced from the databases in Ansys Granta software. The damage model parameter values from the literature are shown in the appendix. Four of the parameters were determined through calibration using fracture test data.

Results and Discussion

The fracture tests were carried out for all the specimens, and the resulting load–displacement plots are shown in Fig. 6. As this figure shows, fracture tests were performed on five identical base metal specimens. The average strength from all the base metal fracture tests is around 740.1 N, and the maximum and minimum strength difference is less than 5%. Only one specimen was used for the fracture strength tests in the nugget zone. The values of the strengths were 404.1, 277.6, and 186.4 N for the specimens with notch lengths 0.35, 0.45, and 0.5 times the width (2.5 mm), respectively.

The five repeated fracture tests for the base metal gauges the repeatability of the proposed fracture test design. Since the deviation in maximum strengths from these five tests is only 5%, the repeatability can be deemed good. Therefore, the choice to perform fracture tests in the nugget zone only once for each notch length is justified. A few observations can be made by comparing the strengths of different fracture specimens. The fracture strengths in the nugget zone decrease for longer notches. This is a predictable observation because the bending stresses are larger for longer notch lengths. A comparison of the material fracture properties between the nugget

Fig. 6 The simulation model of the sample used for the SENB test

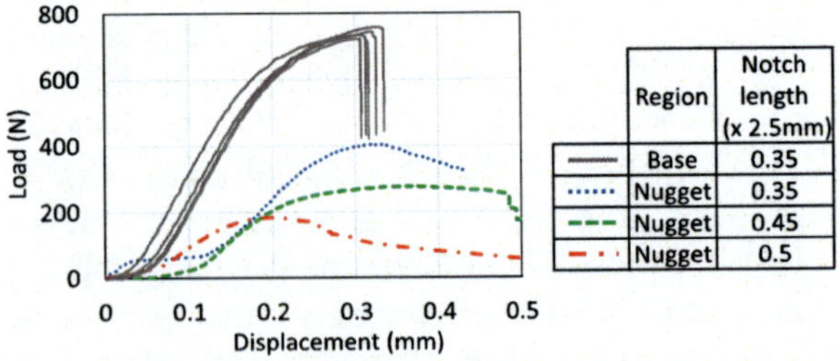

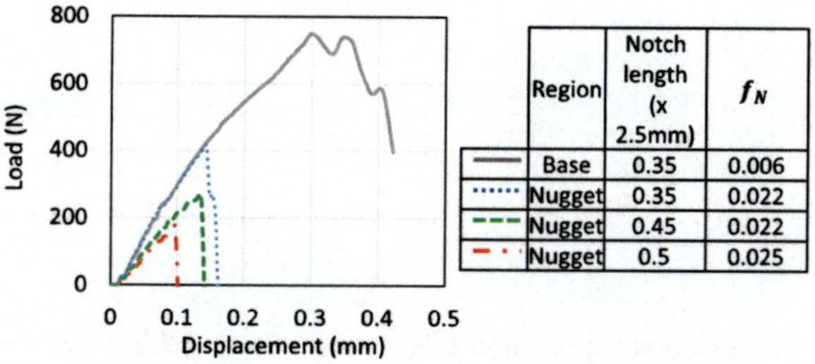

Fig. 7 Simulated load–displacement plots of the SENB tests for various values of a the fraction of nucleating voids (f_N)

zone and the base metal can also be made. For a nugget notch of the same length as in the base metal specimens, i.e., 0.35 times the specimen width, the fracture strength is 404.1 N. This means the base metal's resistance to fracture is approximately 1.8 times larger than that of the nugget zone.

In this paper, four of the GTN model parameters, the fraction of nucleating voids (f_N), mean nucleation strain (ε_N), critical value of the void volume fraction (f_c), and void volume fraction at material failure (f_F) were adjusted so that the simulation strength matched the fracture tests. Figure 7 is illustrates the calibration procedure to determine the parameter values. Here, the values for three of the parameters, ε_N, f_c, & f_F were found to be 0.002, 0.022, and 0.024, respectively. These same values apply to the simulation of all the fracture strengths. As Fig. 7 shows, only the value for one parameter, f_N, needed to be adjusted to match all the strengths. f_N was found to be 0.006 in the base metal, and in the nugget zone, the values were 0.022, 0.022, and 0.025. The larger value in the nugget zone is for the longest notch length.

The choice of assumed parameter values and calibration procedure is validated by successfully recreating every fracture test's strength. In addition, the calibrated parameter values assist in comparing the fracture properties at different depths in the nugget zone. Such a comparison is aided by the choice to vary only one variable, f_N. f_N value at all depths in the nugget zone is approximately the same. Only at the highest depth, f_N value at 0.025 is slightly greater. This implies that fracture properties within the nugget region are constant. This same observation was made in previous studies [17].

Effect of Fracture Properties on Lap Joint Strength

Lap Joint Strength Test Procedure

The same lap joint described previously and shown in Fig. 1a is used for strength testing. It is just that instead of the entire coupon, samples of 25 mm width, as shown in Fig. 8, were cut out. The cutout section was sufficiently far from the ends of the weld. As a result, the analysis can assume a 2D plane stress behavior in the weld direction. This sample was then fixed onto a uniaxial testing machine using clamps at

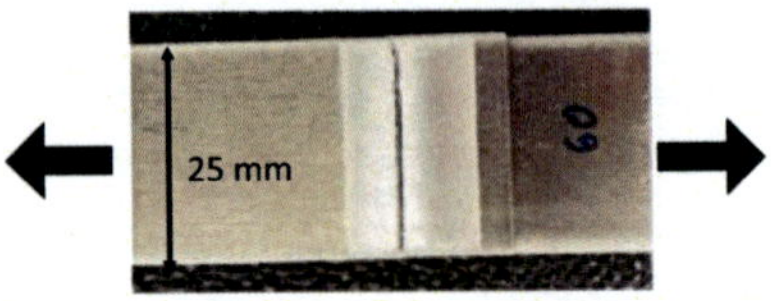

Fig. 8 Top view of the FSLW sample used for lap shear strength testing. The fracture location after the test is also shown

both ends. Then, in a displacement-controlled quasistatic mode, the machine applied a tensile load normal to the weld direction until joint failure. In such a loading case, it can be said that the joint was shear loaded locally at the interface between the sheets, i.e., a lap shear strength test was performed.

Lap Joint Strength Simulation Setup

Lap joint strength simulation uses a procedure similar to the fracture test simulation in the previous section. As shown in Fig. 9, a 2D FEA model of the cross-section geometry of the 3-sheet lap joint was built. The 3-sheets were individually modeled and meshed. The joint was then modeled using tie constraints at the joined interfaces between the sheets. Tie constraints prevent relative motion between the two surfaces, mimicking the joint. The dimensions of the joined surface were from the micrographs in Fig. 1a. The FEA mesh for the top two AA7055 sheets is such that it is the finest (0.1 mm-sided square elements) in the regions closest to the stir zone. In the rest of the sheet, the horizontal dimension of the elements was increased to 0.4 mm. The bottommost sheet used unstructured mesh with both 3-noded and 4-noded elements of approximately 0.4 mm side dimension. Only near the interface of the joint, an element dimension of 0.1 mm was specified.

Parameters of the GTN model were obtained from the calibration study in the previous section. For the region approximating the stir zone, the value of the GTN model parameter f_N was set to 0.022, and for the rest of the region corresponding to the base metal, f_N was set to 0.006. No damage model was used for the bottommost AA6022 sheet because fracture was not experimentally observed here. An extra simulation was also built, assuming the entire top two sheets of the lap joint be made up of just the AA7055-T6 base metal. Such simulation models were tensile loaded,

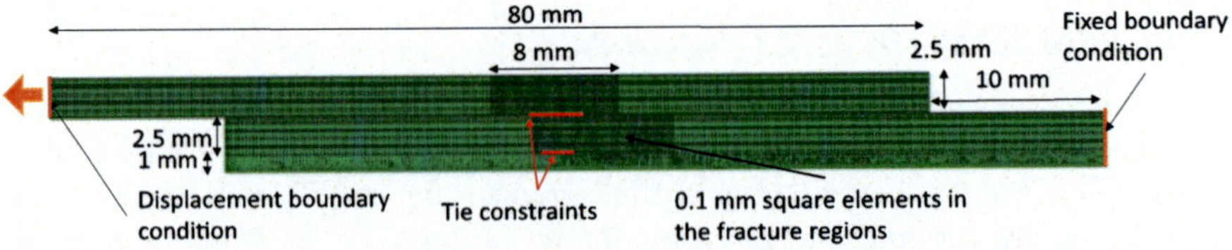

Fig. 9 2D FEA simulation model of the sample used for the lap shear strength test

using displacement-controlled boundary conditions on the ends of the test specimen. The simulations were carried out until the joint failed.

Results and Discussion

The load–displacement plots for lap shear from both experimental testing and simulation are shown in Fig. 10. Experimentally, the FSLW joint almost approaches its maximum strength at initial fracture itself (~0.3 mm displacement) but fails only after some more displacement. The fracture happened in the nugget zone of the top sheet, as shown in Fig. 8. The crack started at the end of the joined interface between the top two sheets and then extended in the direction normal to the loading. The experimentally determined maximum load carried per unit weld length was 585 N/mm. The simulation predicted strength was 1178 N/mm when assuming the fracture properties of the whole joint to be the same as the base metal. When the calibrated fracture properties were used for the nugget region, the simulation predicted strength was 633 N/mm. The fracture mode in the simulations was the same mode observed in the experiment. Note that simulations predict a steep loss in the load right after the point of initial fracture.

The simulation using the nugget zone fracture properties predicts the experimental strength with an error of around 8%. The only noticeable discrepancy in the prediction is the displacement to failure. It could be arising from the fact that the 3D nature of the lap joint was ignored in simulations. However, strength is the focus of this study, and some further observations can be made based on excellent strength predictions. Foremost, many approximations made in both the testing and modeling part of this study are justified. In designing the fracture test, it was assumed there is primarily a mode-I type crack. It means the influence of mode II type crack in the mixed mode crack that is expected to be present [14] is low. The end of the joint interface between the top sheets acted essentially as a crack tip from which a mode-I crack originated in the direction normal to the load. Another justified assumption would be neglecting other aspects of the joint microstructure, such as voids, oxide line shapes,

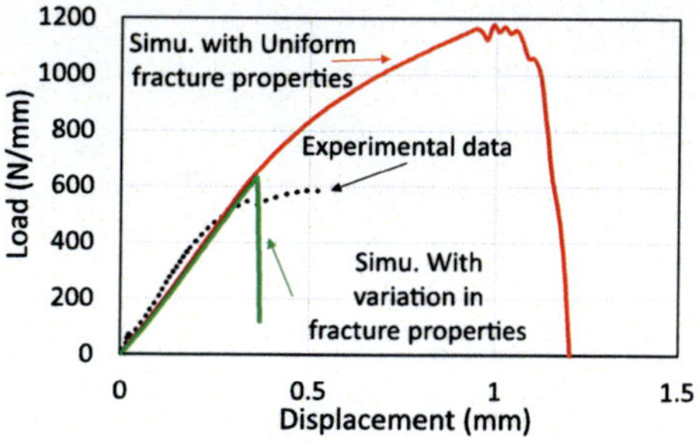

Fig. 10 The load–displacement plots of the lap shear strength test from the experiment and the simulations

and hardness changes near the weld. It also implies that for this AA7055 FSLW joint, the influence of these microstructural aspects on the strength must be insignificant.

Another observation from our study is regarding the research direction to be taken if increasing the FSLW joint strength is an objective. The strength was predicted to be almost 2 times higher in the simulation that assumed base metal fracture properties. Such a higher strength allows us to speculate that process modification could potentially result in a FSLW joint that is significantly stronger.

Conclusions

In this study, simulations and experiments were performed to determine the role of fracture properties on the lap shear strength of a three-sheet (AA7055-7055–6022) FSLW joint. Firstly, specially designed fracture tests were carried out to investigate the fracture strengths within the friction stir weld and determine the fracture parameters for FEA simulations. We found that the simulation with measured weld fracture properties accurately predicted the lap joint strength. The study concludes that fracture property within the nugget zone is the most critical factor determining the strength of AA7055 FSLW joints. In addition, the accuracy of the simulation validates the procedures proposed in this paper to investigate fracture properties in FSW.

Based on the knowledge of the importance of fracture properties, future work would investigate the fracture properties in various other weld regions, not just the nugget zone. Additionally, an understanding of the relationship between fracture properties and FSW process parameters will be developed to improve the strength of FSLW joints.

Acknowledgements The authors are thankful for funding from the US Department of Energy's Vehicle Technology Office under the Joining Core Program 2.0. The authors are also thankful to Honda R&D Inc and Arconic Inc. for providing support in modeling setup and materials needed for experimental work.

Appendix

Gurson-Tvergard-Needleman (GTN) Damage Model

The yield condition in the Gurson-Tvergard-Needleman (GTN) damage model [24] is given as

$$\Phi = \left(\frac{q}{\sigma_y}\right)^2 + 2q_1 f^* \cosh\left(-q_2\frac{3p}{2\sigma_y}\right) - \left(1 + q_3 f^{*2}\right) = 0, \qquad \text{(A1)}$$

where q is the effective mises stress, p is the hydrostatic pressure, σ_y is the yield stress of the fully dense matrix material as a function of the equivalent plastic strain, and $q_1, q_2, \& q_3$ are material parameters. $f^*(f)$ is a function of the void volume fraction f used to model the loss of stress-carrying capacity accompanying void coalescence, and it is defined as

$$f^* = \begin{cases} f & \text{if} f \le f_c \\ f_c + \frac{\overline{f}_F - f_c}{f_F - f_c}(f - f_c) & \text{if} f_c < f < f_F \\ \overline{f}_F & \text{if} f \ge f_F \end{cases}, \tag{A2}$$

where $\overline{f}_F = \frac{q_1 + \sqrt{q_1^2 - q_3}}{q_3}$. In Eq. A2, f_F is the void volume fraction beyond which the material fails to carry any load and f_c is the critical value of the void volume fraction. The change in void volume fraction $\dot{f}$ is

$$\dot{f} = \dot{f}_N + \dot{f}_G, \tag{A3}$$

where $\dot{f}_N$ is the change due to void nucleation and $\dot{f}_G$ is the change due to void growth. These two kinds of change in void volume fraction are defined as

$$\begin{aligned} \dot{f}_G &= (1 - f)\sum_{k=1}^{3} \dot{\varepsilon}_{kk}^{pl}, \\ \dot{f}_N &= A\dot{\varepsilon}_{\text{eff}}^{\text{pl}}, \end{aligned} \tag{A4}$$

where $\dot{\varepsilon}_{\text{kk}}^{\text{pl}}$ denotes the diagonal components of the plastic strain rate matrix and $\dot{\varepsilon}_{\text{eff}}^{\text{pl}}$ denotes the effective plastic strain rate. A is a proportionality constant that is defined as

$$A = \frac{f_N}{s_N\sqrt{2\pi}} \exp\left(-\frac{1}{2}\left(\frac{\varepsilon_{\text{eff}}^p - \varepsilon_N}{s_N}\right)^2\right), \tag{A5}$$

where f_N is the volume fraction of the nucleating voids, ε_N is the mean nucleation strain and s_N is the standard deviation of the nucleation strain. All these three constants are material parameters.

In total, the GTN model has nine material parameters. Some of these parameter values were taken from the literature. For instance, $q_1, q_2, \& q_3$ were set to be 1.5, 1 & 2.25, respectively [25]. The initial void volume fraction f is assumed to be zero, and s_N is assumed to be equal to ε_N. The rest of the parameters, $f_N, \varepsilon_N, f_c$and f_F are to be determined from calibration to the test data.

References

1. Long RS, Boettcher E, Crawford D (2017) Current and future uses of aluminum in the automotive industry. JOM 69(12):2635–2639. https://doi.org/10.1007/s11837-017-2554-9
2. Cao J, Banu M (2020) Opportunities and challenges in metal forming for lightweighting: review and future work. J Manuf Sci Eng Trans ASME 142(11):1–24. https://doi.org/10.1115/1.4047732
3. Threadgilll PL, Leonard AJ, Shercliff HR, Withers PJ (2009) Friction stir welding of aluminium alloys. Int Mater Rev 54(2):49–93. https://doi.org/10.1179/174328009X411136
4. Mishra RS, De PS, Kumar N (2014) Friction stir welding and processing: science and engineering. Springer International Publishing, Cham
5. Shrivastava A, Overcash M, Pfefferkorn FE (2015) Prediction of unit process life cycle inventory (UPLCI) energy consumption in a friction stir weld. J Manuf Process 18:46–54. https://doi.org/10.1016/j.jmapro.2014.10.006
6. Ma ZY, Feng AH, Chen DL, Shen J (2018) Recent advances in friction stir welding/processing of aluminum alloys: microstructural evolution and mechanical properties. Crit Rev Solid State Mater Sci 43(4):269–333. https://doi.org/10.1080/10408436.2017.1358145
7. Hua-Bin C, Yan K, Lin T, Shan-Ben C, Cheng-Yu J, Zhao Y (2006) The investigation of typical welding defects for 5456 aluminum alloy friction stir welds. Mater Sci Eng A 433(1–2):64–69. https://doi.org/10.1016/j.msea.2006.06.056
8. Zhang J, Upadhyay P, Hovanski Y, Field DP (2018) High-speed friction stir welding of AA7075-T6 sheet: microstructure, mechanical properties, micro-texture, and thermal history. Metall Mater Trans A Phys Metall Mater Sci 49(1):210–222. https://doi.org/10.1007/s11661-017-4411-4
9. Nielsen KL (2008) Ductile damage development in friction stir welded aluminum (AA2024) joints. Eng Fract Mech 75(10):2795–2811. https://doi.org/10.1016/j.engfracmech.2008.01.012
10. Liu HJ, Fujii H, Maeda M, Nogi K (2003) Tensile properties and fracture locations of friction-stir-welded joints of 2017–T351 aluminum alloy. J Mater Process Technol 142(3):692–696. https://doi.org/10.1016/S0924-0136(03)00806-9
11. Yazdanian S, Chen ZW, Littlefair G (2012) Effects of friction stir lap welding parameters on weld features on advancing side and fracture strength of AA6060-T5 welds. J Mater Sci 47(3):1251–1261. https://doi.org/10.1007/s10853-011-5747-6
12. Balakrishnan M, Leitão C, Arruti E, Aldanondo E, Rodrigues DM (2018) Influence of pin imperfections on the tensile and fatigue behaviour of AA 7075–T6 friction stir lap welds. Int J Adv Manuf Technol 97(5–8):3129–3139. https://doi.org/10.1007/s00170-018-2172-x
13. Yadava MK, Mishra RS, Chen YL, Carlson B, Grant GJ (2010) Study of friction stir joining of thin aluminium sheets in lap joint configuration. Sci Technol Weld Join 15(1):70–75. https://doi.org/10.1179/136217109X12537145658733
14. Yang C, Tomblin JS, Guan Z (2003) Analytical modeling of ASTM lap shear adhesive specimens. In: U.S. department of transportation federal aviation administration office of aviation research, Washington, DC. Report No. DOT/FAA/AR-02/130, p 78
15. Zhang S (1999) Stress intensities derived from stresses around a spot weld. Int J Fract 99(4):239–257. https://doi.org/10.1023/A:1018608615567
16. Sevim I (2006) Effect of hardness to fracture toughness for spot welded steel sheets. Mater Des 27(1):21–30. https://doi.org/10.1016/j.matdes.2004.09.008
17. Derry CG (2008) Characterisation and modelling of toughness in aerospace aluminium alloy friction stir welds
18. Kulekci MK, Sevim I, Esme U (2012) Fracture toughness of friction stir-welded lap joints of aluminum alloys. J Mater Eng Perform 21(7):1260–1265. https://doi.org/10.1007/s11665-011-0017-y
19. Strombeck A, dos Santos J, Torster F, Laureano P, Kocak M (1999) Fracture toughness behaviour of FSW joints on aluminium alloys. In: Proceedings of the first international symposium on friction stir welding, pp 1–11

20. Pirondi A, Fersini D (2009) Simulation of ductile crack growth in thin panels using the crack tip opening angle. Eng Fract Mech 76(1):88–100. https://doi.org/10.1016/j.engfracmech.2008.05.008
21. Gallais C et al (2007) Multiscale analysis of the strength and ductility of AA 6056 aluminum friction stir welds. Metall Mater Trans A 38(5):964–981. https://doi.org/10.1007/s11661-007-9121-x
22. Derry CG, Robson JD (2008) Characterisation and modelling of toughness in 6013–T6 aerospace aluminium alloy friction stir welds. Mater Sci Eng A 490(1–2):328–334. https://doi.org/10.1016/j.msea.2008.01.044
23. Nielsen KL, Pardoen T, Tvergaard V, de Meester B, Simar A (2010) Modelling of plastic flow localisation and damage development in friction stir welded 6005A aluminium alloy using physics based strain hardening law. Int J Solids Struct 47(18–19):2359–2370. https://doi.org/10.1016/j.ijsolstr.2010.03.019
24. Tvergaard V (1981) Influence of voids on shear band instabilities under plane strain conditions. Int J Fract 17(4):389–407. https://doi.org/10.1007/BF00036191
25. Aravas N (1987) On the numerical integration of a class of pressure-dependent plasticity models. Int J Numer Methods Eng 24(7):1395–1416. https://doi.org/10.1002/nme.1620240713

Part VII
Dissimilar and Non-ferrous

High Speed Butt Joining of 1″ Thick 2139-T8

Hrishikesh Das, Piyush Upadhyay, Reza E. Rabby, Uchechi Okeke, and Martin McDonnell

Abstract Thick plate (≥1″) butt joining of Al alloys is challenging due to high tool forces, uneven material flow, and heat distribution in the through thickness direction. Tool design and welding parameters used for executing thick plate butt joining have remained mostly static over the past decade. Reported welding speed that produces viable joint strength (mostly on 6xxx alloys) typically ranges below 100 mm/min (4 inches/min). Researchers at Pacific Northwest National Laboratory (PNNL) in association with Ground Vehicle Systems Center (US Army GVSC) have been working to demonstrate greater joining efficiency at higher welding speeds in 1″ thick AA2139-T8 plate. Using innovative tool features and effective force and temperature control, a joint efficiency of 80% was demonstrated at a welding speed of 7 inches/min (178 mm/min).

Keywords Thick plate · FSW · High speed FSW · X forces

Introduction

Aluminium (Al) alloy AA2139-T8 is an age-hardened Al–Cu–Mg–Ag alloy. The temper (T8) indicates that the alloy is solutionized, quenched, cold-worked, and artificially aged. The magnesium (Mg) and sliver (Ag) alloying elements yield the metastable Ω precipitate, which increases both strength and ductility [1–4]. The AA2139 alloy has excellent ballistic performance and hence been registered as armor plate by the United States Department of Defense in monolithic form [1, 2]. Ability to effectively butt join sections can enable light weighting of armored tanks. Welding of this alloy is particularly challenging irrespective of the fusion and solid-state joining

H. Das · P. Upadhyay (✉) · R. E. Rabby
Pacific Northwest National Laboratory, 902 Battelle Blvd., Richland, WA 99352, USA
e-mail: piyush.upadhyay@pnnl.gov

U. Okeke · M. McDonnell
DEVCOM Ground Vehicle Systems Center, 6305 E Eleven Mile Rd., Warren, MI 48092, USA

Y. Hovanski et al. (eds.), *Friction Stir Welding and Processing XII*,
The Minerals, Metals & Materials Series,
https://doi.org/10.1007/978-3-031-22661-8_19

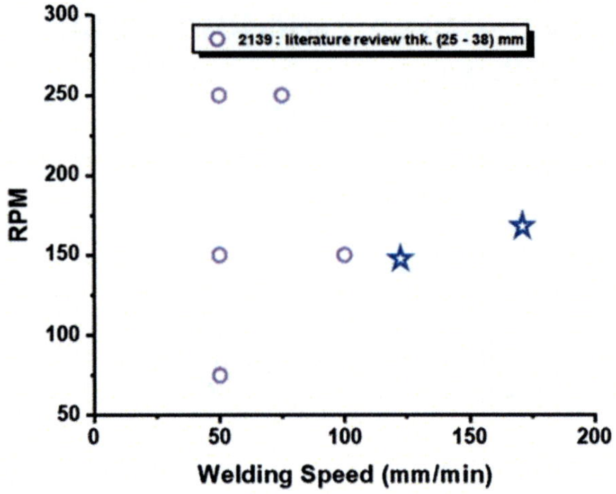

Fig. 1 Welding parameters used in the literature for AA2139 FSW [5, 6]

process. Over the years, a number of works have been published on the microstructure and mechanical property correlation, optimization of process parameters, and effect of tool geometries for friction stir (FS) butt welding of different grades of aluminium alloys. However, only a small percentage of the work consists of joining Al with thickness (≥25 mm). Thick plate joining has direct implications for armored vehicles, ship building industry, cryo-tanks, oil and gas, etc. A summary of the available welding parameters (RPM and welding speed) with AA2139 (≥25 mm) in the literature is shown in Fig. 1. Thompson et al. [5] investigated single and double pass FS welded 25 mm and 38 mm thick AA2139 and reported a maximum joint strength of 379 MPa (75% of joint efficiency). McWilliams et al. [6] reported their research work on micro-hardness based constitutive model development for different FS weld zone (SZ and TMAZ) with validation of experimental results to characterize the mechanical properties of FS welded AA2139-T8 plates (>25 mm). Planar forces are a primary limiting factor for thick-section welding.

Experimental Method

Friction stir butt welding was performed on 1" (25.4 mm) thick AA2139-T8 (ultimate tensile strength = 503 MPa, 4 samples tested). Two 4″ wide plates were clamped together in a butt joint configuration. The FSW tool was made of H13 tool steel and had a 35 mm shoulder diameter, with a single scroll. The tool probe shape was a truncated cone (8-degree taper) with threads and three flats. The probe was 23.6 mm long with 1.4 mm shoulder to pin transition length. Maximum pin diameter was 19 mm (Fig. 2). This tool design was designated as "PN1". Tool design iterations involving opposing thread direction were also fabricated and used for the study. Two of these tools are designated as PN2 and PN3 tools. Both of these tools had cascading opposing pin threads, as seen in Fig. 2. This pin consisted of finer threads with four separate thread regions. The 1st, 3rd, and 4th region contained threads in same direction as the base tool while the thread direction was reversed for the 2nd region. This resulted in two local material flow reversals regions due to opposing thread directions. Z-force-RPM control method was used to make welds reported in this paper. Force experienced by the tool in longitudinal direction (X force) and that in transverse direction (Y force) was also continuously monitored and recorded. Room temperature tensile testing was performed on three specimens from each joining trial at an extension rate of 2.54 mm/min using a 222 kN MTS test frame. The tools were instrumented with type-k thermocouples welded at the shoulder for the purpose for monitoring temperature. A total of 4 welding cases are discussed in this paper. Table 1 below shows the welding conditions discussed in Sect. 3. Securely clamping the two plates to withstand planar forces is a frequent problem in thick plate welding. Separation of the plates can occur due to excessive force or suboptimal clamping. A robust clamping scheme (Fig. 3a) was used to avoid any separation. A typical separation of plates during plunging and resulting tunnel defect (line defect) is shown in Fig. 3b. Proper clamping can reduce or eliminate surface defects, as shown in Fig. 3c.

Fig. 2 Left: Base tool used in this study referred as PN1. Center: First iteration of opposing thread tooling PN2. Right: PN3 tool containing opposing/finer thread. PN1 and PN3 are compared in the Results and Discussion

Table 1 Welding parameters used to run four welding runs discussed in below

Weld #	Tool	Rotational speed (RPM)	Welding speed inches/minutes (mm/min)	Z-force kN
A	PN1	140	3 (76)	71
B	PN3	140	3 (76)	51
C	PN1	165	5 (127)	58
D	PN3	150	7 (178)	76

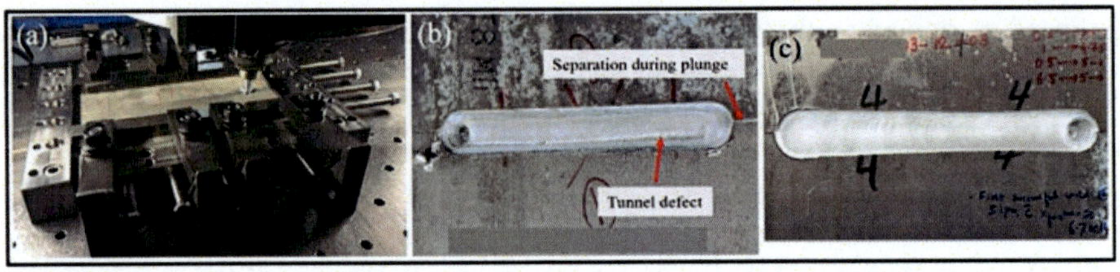

Fig. 3 **a** Clamping setup used for this study, **b** plate separation during plunge led to a tunnel defect, **c** FSW butt weld with no defect made after adequate clamping

Results and Discussion

Planar forces for butt welds made at a welding speed of 76.2 mm/min (3 inches/min) using tool design PN1 and PN3 are compared in Fig. 4. For both the cases, the shoulder temperature (not shown) ranged between 470 and 490 °C. The x-force for the PN3 tool is half that of PN1 (18 kN for PN1 versus 9 kN for PN3). Relative values of the y-force on the other hand were flipped. PN3 generated a y-force approximately 40% higher than that for PN1. The opposing thread feature in the pin of PN3 had a significant effect on the planar forces. While the resultant force for both tools was similar, the force in the traverse direction appeared to be significantly reduced by the reversing pin thread direction.

One of the challenges towards enabling higher welding speeds for thick plate FSW is a concomitant increase in planar forces, especially the x-force. Since an excessive x-force can result in pin shearing and/or machine stalling (due to soft limits imposed to avoid overloading), it is desirable to reduce the x-force during welding. Encouraged by results discussed above, a welding campaign was performed with the PN1 and PN3 tools at progressively higher welding speeds. Welding speed in excess of 5 INCHES/MIN with PN1 resulted in x-force nearing the machine soft limit of 32 kN

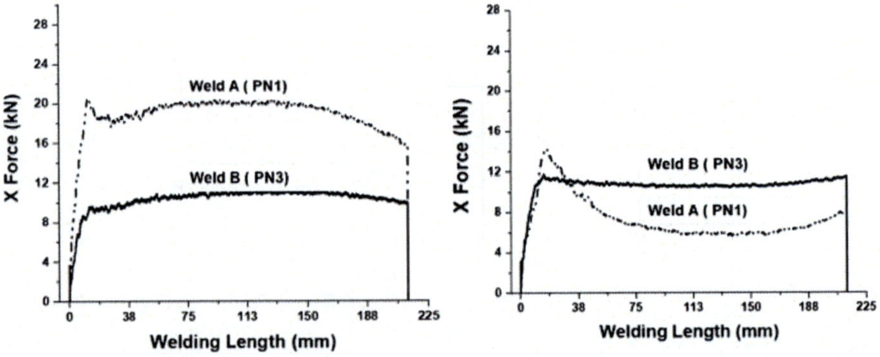

Fig. 4 X-force and y-force plotted against the weld length for weld A (tool PN1) and B (tool PN3). Both were made at a welding speed of 76.2 mm/min (3 inches/min) at 140 RPM

(7.2 klbf). However, similar welds with PN3 experienced much lower x-force. After a few pin design modifications including thread and flat depth, the PN3 tool was able to produce a defect free, high strength weld up to a welding speed of 7 inches/min (178 mm/min).

The weld process responses and joint strengths were compared for two cases: Weld (C) PN1-5 inches/min and (2) PN3-7 inches/min. Figure 5 shows the z-force and measured shoulder temperature evolution during welding the two cases. Weld D required higher z-force and ran about 25 °C hotter at the shoulder compared to weld C. X- and y-forces for the two cases for the length of the weld are plotted in Fig. 6. A trend similar to what was shown in Fig. 4 is apparent. The x-force for Weld D is significantly lower than that for Weld C despite a 40% increase in welding speed. The reduction in x-force is balanced by an increase in y-force. However, the y-force is not stable. There is a peak in the y-force midway along the length of the weld. Further studies with longer weld lengths are suggested for future work to understand the y-force observation in Weld D. Nevertheless, the PN3 tool produced more balanced x- and y-forces compared to the conventional PN1 tool design.

Figure 7 shows the etched macrostructure in the transverse cross section for the two weldments. In both weldments, no defects were observed. The PN1 tool exhibited a traditional, basin-shaped, FS weld profile. However, step features can be seen in the macrograph for the PN3 tool. These "steps" are where the opposing threads meet and can be compared to the physical profile of the tool, as shown in Fig. 2. The stress–strain curves for the weldments (5 inches/min with PN1 and 7 inches/min with PN3) are shown in Fig. 8. The samples welded at 7 inches/min with PN3 had an ultimate tensile strength (UTS) of 400 ± 1 MPa and a 80% joint efficiency of 80%. The samples welded at 5 inches/min with PN1 had an UTS of 406 ± 4 MPa and a joint efficiency of 81%. Additionally strain to failure between the two cases is also similar. Despite a 40% increase in the welding speed for the PN3 tool, the joint strengths and ductility are similar. This suggests that the opposing threads in the PN3

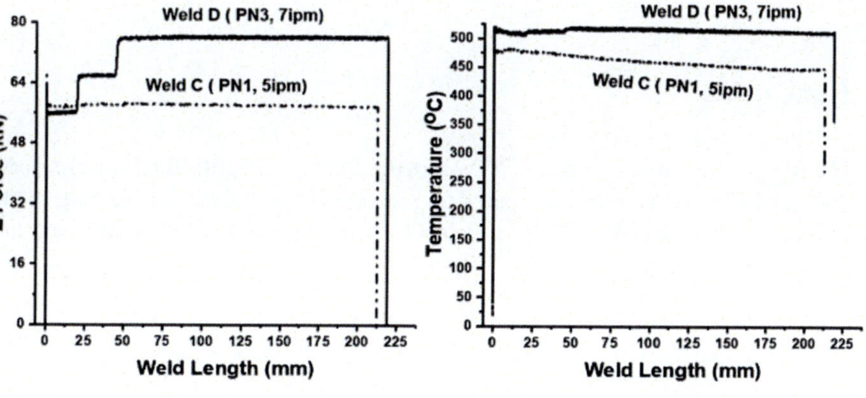

Fig. 5 Left: Z-force versus weld length. Right: measured shoulder temperature versus weld length for Weld C and D

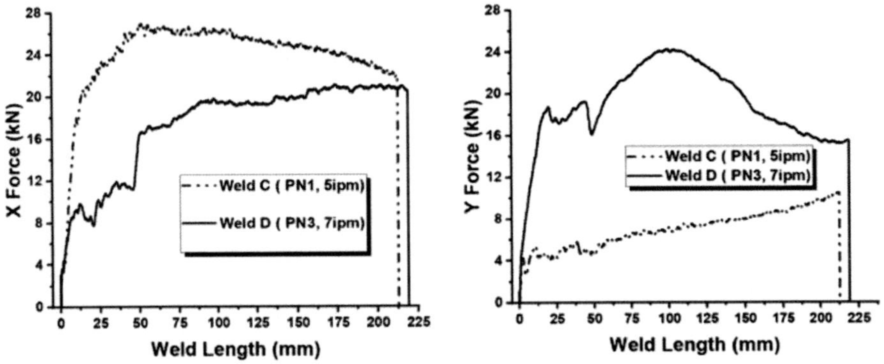

Fig. 6 X-force (left) and y-force (right) plotted against the weld length for Weld C (PN1 tool at 127 mm/min (5 inches/min) and Weld D (PN3 tool at 178 mm/min (7 inches/min)

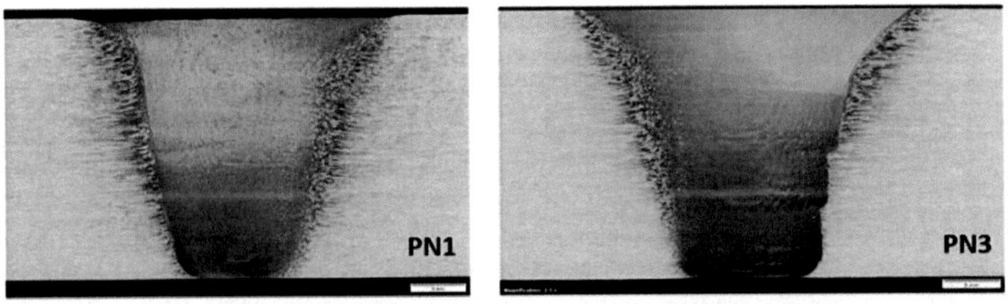

Fig. 7 X-force (left) and y-force (right) plotted against the weld length for Weld C (PN1 tool at 127 mm/min (5 inches/min) and Weld D (PN3 tool at 178 mm/min (7 inches/min)

tool pin allowed for increased welding speeds without significantly impacting the weld strength.

Conclusion

Use of opposing thread design in FSW tool pin showed a significantly lower x-force in 1″ thick 2139 butt welding compared to base tool. This enabled butt joining of plates up to a welding speed of 178 mm/min (7 inches/min) with 80% joining efficiency and no weld defects.

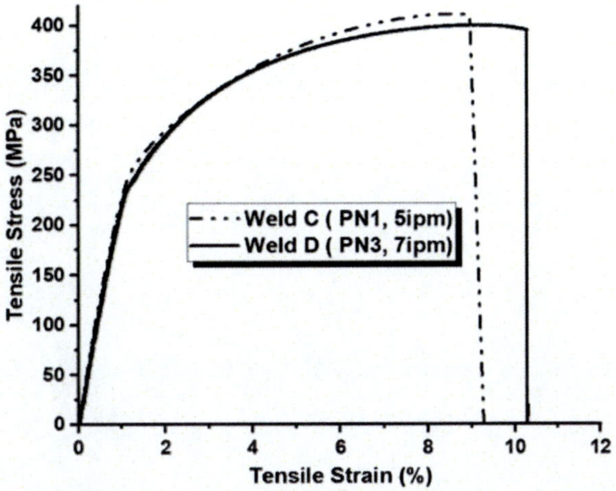

Fig. 8 Tensile stress versus tensile strain plot for joints made at the highest welding speed possible for Case 1 (PN1 tool at 127 mm/min (5 inches/min) and Case 2 (PN3 tool at 178 mm/min (7 inches/min). Note that base metal UTS = 503 MPa

Acknowledgements The U.S. Army Tank Automotive Research, Development, and Engineering Center supported this work [Military Interdepartmental Purchase Request 11613546, 11079043]. Pacific Northwest National Laboratory (PNNL) is operated by the Battelle Memorial Institute for the United States Department of Energy under contract DE-AC06-76LO1830. The authors also thank PNNL staff Timothy Roosendaal, and Anthony Guzman for assisting in metallography and mechanical testing.

References

1. Knowles KM, Stobbs WM (1988) The structure of 111 age-hardening precipitates in Al–Cu–Mg–Ag alloys. Acta Crystallogr B 44:207–227. https://doi.org/10.1107/S0108768187012308
2. Muddle BC, Polmear IJ (1989) The precipitate Ω phase in Al–Cu–Mg–Ag alloys. Acta Metall 37:777–789. https://doi.org/10.1016/0001-6160(89)90005-9
3. Grujicic M, Arakere G, Yen C-F, Cheeseman BA (2011) Computational investigation of hardness evolution during friction-stir welding of AA5083 and AA2139 aluminium alloys. J Mater Eng Perform 20(7):2011–1097. https://doi.org/10.1007/s11665-010-9741-y
4. Thompson B, Doherty K, Niese C, Eff M, Stotler T, Pramann Z, Seaman J, Spencer R, White P (15–17 May 2012) Friction stir welding of thick section aluminium for military vehicle applications. In: 9th International symposium on friction stir welding (9ISFSW). Huntsville, AL
5. McWilliams BA, Yu JH, Yen C-F (2013) Numerical simulation and experimental characterization of friction stir welding on thick aluminium alloy AA2139-T8 plates. Mater Sci Eng, A 585:243–252
6. Hornbuckle BC, Murdoch HA, Roberts AJ, Kecskes LJ, Tschopp MA, Doherty KJ, Yu JH, Darling KA (2017) Property mapping of friction stir welded Al-2139 T8 plate using site specific shear punch testing. Mater Sci Eng, A 682:192–201. https://doi.org/10.1016/j.msea.2016.11.032

Joining Cast Mg AZ91 and Wrought Al 6082 Through Friction Stir Welding

Krzysztof Mroczka, Stanisław Dymek, Adam Pietras, Aleksandra Węglowska, Carter Hamilton, and Mateusz Kopyściański

Abstract Dissimilar welds of cast magnesium AZ91 with wrought aluminium 6082-T6 were fabricated by friction stir welding (6 mm thick) with the alloys alternately placed on the advancing and retreating sides. The unique weld microstructures were characterized through light and electron microscopies (SEM, EDS), tensile tests were performed on samples containing the entire welds and the microhardness was measured across the weld cross-sections. Preliminary results demonstrate that some regions contain a significant amount of the brittle $Al_{12}Mg_{17}$ phase. To complement the microstructural and mechanical characterization, numerical simulations of the welding process provided deeper insight into the material flow during mixing and to the temperature distributions across the weld zones. Future work will include micro tensile specimens (cross-sectional area of 0.7×0.7 mm) to supplement the conventional tensile tests. These micro tensile specimens will be excised from distinct weld areas to highlight differences in local properties that influence weld behavior.

Keywords Friction stir welding · Dissimilar metals · Magnesium alloys · Aluminium alloys · Characterization

K. Mroczka
Department of Materials Engineering, Faculty of Materials Science and Physics, Cracow University of Technology, Krakow, Poland

S. Dymek · M. Kopyściański
Faculty of Metal Engineering and Industrial Computer Science, AGH University of Science and Technology, Krakow, Poland

A. Pietras · A. Węglowska
The Łukasiewicz Research Network Institute of Welding, Gliwice, Poland

C. Hamilton (✉)
Department of Mechanical and Manufacturing Engineering, Miami University, College of Engineering and Computing, 56N Garland Hall, Oxford, OH 45056, USA
e-mail: hamiltbc@miamioh.edu

Y. Hovanski et al. (eds.), *Friction Stir Welding and Processing XII*,
The Minerals, Metals & Materials Series,
https://doi.org/10.1007/978-3-031-22661-8_20

Introduction

This research explores the joining of cast Mg AZ91 and wrought Al 6082-T6 through FSW. Both alloys are lightweight, structural alloys with good corrosion resistance. Al 6082 is a precipitation-strengthened, age-hardenable alloy, and as such, the processing temperatures during FSW could achieve or exceed the precipitation/dissolution temperatures of the alloy. The typical precipitation sequence of 6082 is *GP* Zones → β'' (formation at 250 °C) → β' (formation at 300 °C) → β (formation at 475 °C) with dissolution of the β phase and supersaturation of the matrix occurring at 525 °C [1]. Under typical FSW conditions for age-hardenable aluminium alloys, for similar or dissimilar joining, the temperatures within the weld zone can exceed the solution heat treatment temperature(s) of the alloy(s) leading to reprecipitation of the primary strengthening phases (*GP* or *GPB* zones) upon cooling. As a result, the fusion zone experiences a hardness recovery in relation to the hardness on the advancing and retreating sides of the weld and the thermo-mechanically affected zone.

The primary alloy elements of Mg AZ91 are Al (~8%), Zn (~0.7%) and Mn (~0.2%) with the balance of Mg at ~90% [2]. The addition of Zn and Al augments the mechanical performance of the alloy; however, the binary phase diagram for Mg–Al reveals the presence of the brittle intermetallic phase $Mg_{17}Al_{12}$ and its equilibrium conditions with solid solution α-Mg. The eutectic reaction between these phases occurs at ~68% Mg and at a temperature of 437 °C [3]. Given that the maximum solid solubility of Al in Mg is ~10%, the solidus temperature of AZ91 is in proximity to the eutectic temperature at 470 °C. Therefore, for typical FSW conditions and temperatures, liquid formation in AZ91 during processing is possible with the subsequent formation of the brittle eutectic $Mg_{17}Al_{12}$/α-Mg structure upon cooling. The presence of this eutectic phase in the weld zone would be detrimental to its mechanical performance.

Investigations into the joining of Mg alloys with Al alloys through FSW have been performed for many years. McLean et al. examined the feasibility of joining AZ31 to Al 5053 through FSW and found liquation in the Mg alloy occurring within the weld zone [4]. The resulting welds contained the brittle $Mg_{17}Al_{12}$/α-Mg structure as a divorced lamellar eutectic. Gerlich et al. studied friction stir spot welding as applied to sheets of AZ91 with Al 6111 and the peak processing temperatures achieved [5]. Regardless of which alloy was placed on top or bottom, processing temperatures in areas around the tool would reach or exceed the solidus and eutectic temperatures of AZ91, and the brittle eutectic phase would form. Similarly, Buffa et al. found the eutectic structure in the friction stir welds of AZ31 with Al 6016 [6]. Though the eutectic structure hampered the mechanical performance of the welds, Buffa determined that sound joints could be produced only if the Mg alloy were placed on the advancing side. In studying the FSW of AZ 91 and Al 6082, Sameer et al. also concluded that sound joints could be achieved with the Mg alloy on the advancing side; however, the welds were still characterized by the presence of the eutectic structure [7].

The matter of which alloy to place on the advancing or retreating sides remains a point of debate as various research efforts have shown that superior weld quality in dissimilar welds is obtained when the softer alloy is placed on the advancing side and the harder alloy on the retreating side and vice versa [8–11]. Two primary challenges to the FSW of Mg alloys with Al alloys, therefore, are to determine which alloy position maximizes joint quality/performance and to determine the optimal processing methods to minimize or eliminate the eutectic structure in the weld zone. This paper presents the preliminary results from the investigation of Mg AZ91-Al 6082 welds with highlights and details on the next steps to be taken to address these challenges.

Experimental Procedure

Friction Stir Welding

Plates of cast Mg AZ91 and Al 6082-T6 with dimensions of 100 × 250 × 6 mm were obtained for friction stir welding at the Institute of Welding in Gliwice, Poland. Prior to joining, the workpiece edges were sanded to remove the oxide layers and then cleaned with solvent to remove contaminants from the weld seam. The goals of this project are to investigate two types of FSW tools for this materials system, a conventional FSW tool with cylindrical/conical pin and a dual-speed FSW tool and to investigate the placement of the alloys alternately on the advancing and retreating sides. At this time, only joining with the conventional tool and with Al 6082 on the advancing side has been accomplished. The conventional tool has a scrolled shoulder with a 14 mm radius and 2.5 mm pitch, and the pin is 5.8 mm long with a 5.5 mm radius and a thread pitch of 3 mm. Joining was performed under force control (30 kN), but the other process parameters, i.e., rotation speed and welding speed, were varied to determine those parameters that produce sound welds for this material system. Investigations with the dual-speed tool and with additional weld configurations will comprise future studies.

Microscopy

Weld microstructures were investigated with light microscopy using a reagent containing $C_6H_2(NO_2)_3OH$ 4.2 g, CH_3COOH 1 ml, H_2O 10 ml, C_2H_5OH 96% 75 ml to highlight the presence of the $Mg_{17}Al_{12}$ intermetallic phase. Scanning electron microscopy with EDS investigated the microstructure and fracture specimens. At this time, only the optical microscopy and SEM work with EDS has been accomplished; however, details of the microstructure will be additionally studied applying TEM microscopy and EBSD in future studies.

Mechanical Testing

At this time, tensile testing and hardness evaluation is incomplete, but will be part of additional investigations. For those future studies, two types of tensile test specimens will be utilized: micro—local properties testing using specimens with a reduced section cross-sectional area of 0.7 × 0.7 mm and macro—global properties testing perpendicular to welding direction using specimens with a reduced section cross-sectional area of 16 × 6 mm. Wire electrical discharge machining will be used to excise the tensile specimens.

Numerical Simulation

A numerical simulation of the friction stir welding process was adapted from the model previously developed by the authors for welding of dissimilar aluminium alloys [12]. The solution approach of the current simulation, including boundary conditions for flow velocities around the tool, flow stress, viscosity, strain rate and temperature, is consistent with the authors' prior simulation as detailed in reference [12].

Results and Discussion

Figure 1A presents the weld microstructure obtained at 710 RPM tool rotation and 90 mm/min weld speed with Al 6082 on the advancing side and Mg AZ91 on the retreating side. With the conventional tool, these parameters produced the highest quality weld thus far manufactured for this research program. The image shows uniform mixing of the alloys throughout the weld zone with interleaving of material layers derived from the workpiece materials.

Closer examination of the optically blue layers in this image, however, reveals that these layers are the $Mg_{17}Al_{12}/\alpha$-Mg eutectic structure, as shown in Fig. 1b. The SEM image in Fig. 2 additionally reveals that the weld zone layers are not just the eutectic structure interleaved with aluminium, but that additional, distinct layers are also present, appearing as lighter and darker contrasting bands. The line scan of the layers also presented in Fig. 2 shows the Mg $K\alpha 1$ and Al $K\alpha 1$ response from each region and highlights the varying composition of each primary element within the layers.

To facilitate the identification of the phases present and the primary constituents in these layers, numerous EDS spectra were taken within the eutectic channel and across the other contrasting bands. As shown in Fig. 3, the location of the exemplar spectra highlighted in the figure are: (A) matrix of the eutectic channel, (B) island

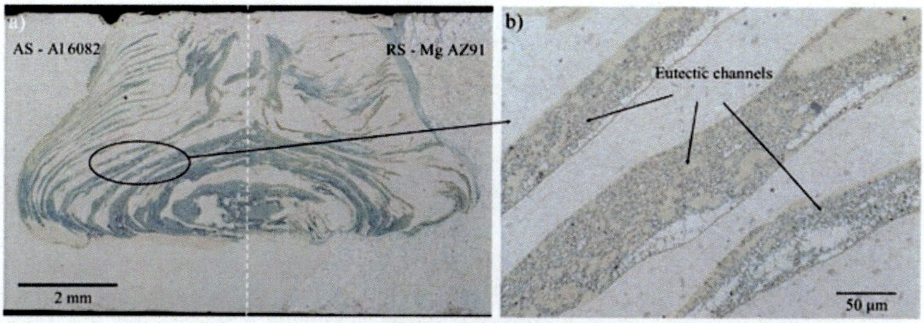

Fig. 1 **a** Weld microstructure at 710 RPM/90 mm/min with Al 6082 on the advancing side and Mg AZ91 on the retreating side and **b** higher magnification image of eutectic channel layers

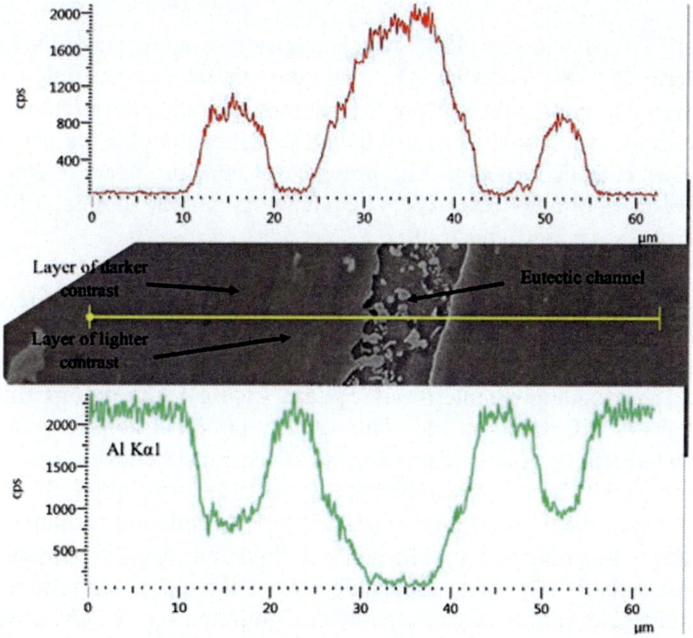

Fig. 2 SEM image of eutectic channels with line scan showing Mg Kα*1* and Al Kα*1* responses

phase within the eutectic channel, (C) lighter contrast layer and (D) darker contrast layer.

The spectral compositions from each of these locations are also presented in Fig. 3. The chemical composition of Spectrum A is 84.1% Mg and 10.7% Al which is close to the composition of the α-Mg solid solution at maximum solid solubility of Al in Mg, but slightly lean in Mg content. The leaner Mg content suggests non-equilibrium solidification of the solid solution at/below the eutectic temperature. Spectrum B taken from an island phase within the eutectic channel reveals a composition of 76.1%

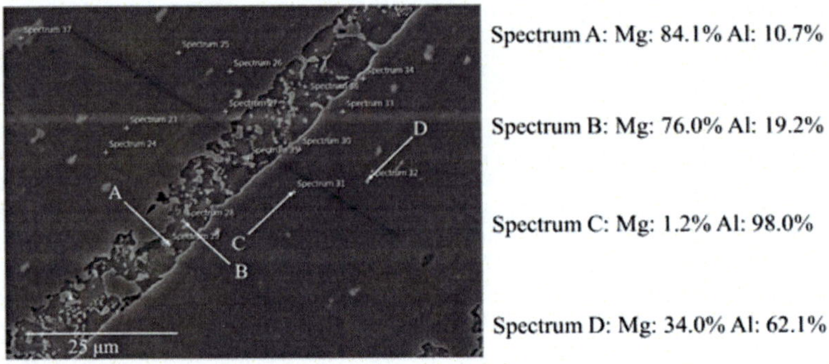

Fig. 3 EDS spectra taken from within and outside the eutectic channel: **a** matrix of the channel, **b** island phase within channel, **c** lighter contrast layer, **d** darker contrast layer

Mg and 19.2% Al, which is close to the anticipated composition of the $Mg_{17}Al_{12}$/α-Mg eutectic, 68% Mg and 19% Al. Here, however, the composition is rich in Mg, but this would be consistent with non-equilibrium solidification at/below the eutectic temperature—the Mg solute rejected by the solidification of α-Mg is manifested as a Mg-richer eutectic structure. The presence of eutectic layers is consistent with the observations of other researchers who have studied the joining of Mg/Al alloys through solid-state processes [5–8].

Spectrum C taken from the layer showing lighter contrast in the SEM images reveals a chemical composition of 1.2% Mg and 98.0%, which is entirely consistent with the Al 6082 composition. Spectrum D taken from the darker contrast layer presents a composition of 34.0% Mg and 62.1% Al. The equilibrium phase diagram for Mg/Al contains another intermetallic phase, Al_3Mg_2, with a composition of ~36% Mg and ~64% Al; therefore, this layer within the weld zone may represent yet another intermetallic phase influencing the mechanical behavior and properties as also observed by Xu et al. when joining Mg AZ31 and Al 5A06 [13].

Temperature distribution predictions from the simulation certainly support the notion that temperatures are sufficient for liquation and the subsequent formation of the eutectic. Figure 4a shows the predicted temperature profile for the 710 RPM/90 mm/min process parameters with aluminium on the advancing side. The temperature cross section is taken behind the tool just past the pin profile, i.e., at 6 mm. The isotherms indicated in the image show the 437 °C eutectic temperature and the 470 °C solidus temperature of α-Mg. A significant portion of the upper weld zone near the surface and under the tool shoulder experiences temperatures above the solidus temperature. In this region, therefore, some melting will occur. Many research studies have shown that during FSW surface material is extruded into the weld zone typically from the advancing side.

Provided the process parameters are suitable for thorough mixing of the workpieces, any liquid formed at the surface would be extruded into the weld zone and interleaved with plasticized solid material from the workpieces. As suggested by the temperature profile in Fig. 4a, the bulk of the weld zone below the surface is above the

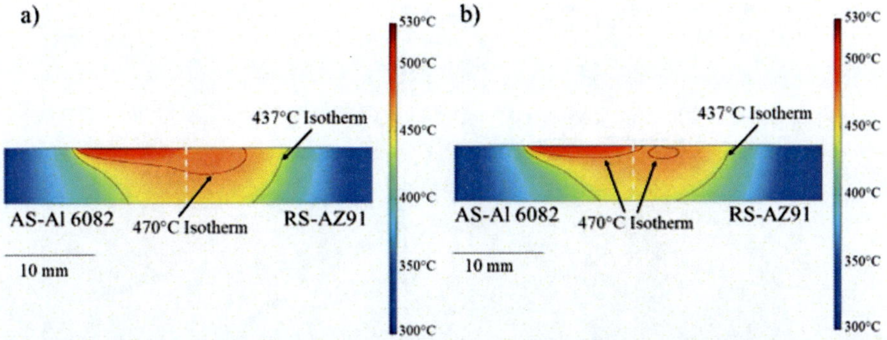

Fig. 4 Predicted temperature distribution for the **a** 710 RPM/90 mm/min and AS Al 6082/RS Mg AZ91 weld condition and **b** 710 RPM/140 mm/min and AS Al 6082/RS Mg AZ91 weld condition

eutectic temperature, so much of the liquid extruded from the surface would remain in that state until cooling occurs, forming the $Mg_{17}Al_{12}$/α-Mg eutectic structure upon solidification.

Reducing the heat input could mitigate the formation of the eutectic structure if the surface temperatures could remain below 470 °C; however, it is important to keep in mind that the solution heat treatment temperature of Al 6082 is 525 °C. Welding temperatures too far below this level may not lead to as effective mixing of the aluminium workpiece as that at higher process temperatures. Figure 5 shows the weld microstructure obtained at 710 RPM tool rotation and 140 mm/min weld speed with the AS Al 6082/RS Mg AZ91 configuration. The faster weld speed reduces the heat input, but as seen in the figure, the reduction in heat input is insufficient to eliminate the eutectic phase. Rather, the eutectic phase dominates the top portion of the weld nugget where the temperatures are the highest, and, overall, the weld shows poor mixing of the aluminium workpiece due to the lower temperatures. Figure 4b shows the predicted temperature profile for these weld conditions taken on a cross section from the same position as Fig. 4a. Though the zone for which the temperature exceeds 470 °C is smaller, there is still the potential for liquation at the surface and extrusion of melt into the weld zone; however, given the poor mixing of the aluminium at these temperatures, the liquation remains confined to the surface area with the eutectic structure developing predominantly in this area upon cooling.

Conclusions

Preliminary studies on joining cast Mg AZ91 with wrought Al 6082-T6 were performed. Characterization work through optical microscopy, SEM and EDS was initiated, and a numerical simulation of the process was created to help glean deeper insight into the temperature distributions across the weld zones. The following observations and conclusions are noted:

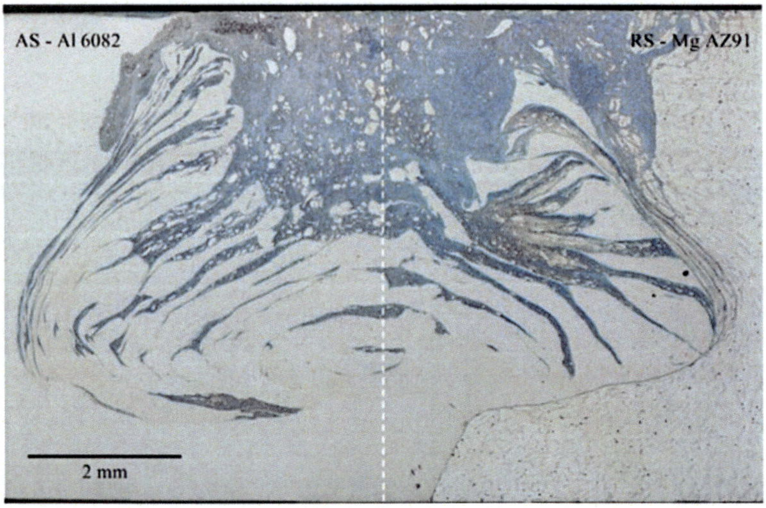

Fig. 5 Weld microstructure at 710 RPM/90 mm/min with Al 6082 on the advancing side and Mg AZ91 on the retreating side

1. The $Mg_{17}Al_{12}/\alpha$-Mg eutectic structure is prevalent in the weld zone. The weld nugget is primarily comprised of interleaved layers of this eutectic structure with workpiece materials.
2. The eutectic channels are comprised of $Mg_{17}Al_{12}/\alpha$-Mg eutectic within an α-Mg matrix. The eutectic is richer in Mg than typical, and the solid solution is leaner in Mg, suggesting non-equilibrium solidification.
3. Other layers comprising the weld nugget include the baseline Al 6082 and potentially the other intermetallic phase common in Mg AZ91, Al_3Mg_2.
4. Temperature predictions from the numerical simulation demonstrate that the upper portion of the weld nugget experience temperatures at or above the solidus temperature of Mg AZ91 and that the bulk of the weld nugget remains at or above the eutectic temperature.

Acknowledgements This work was supported by the Polish National Science Center, Grant No. UMO-2018/31/B/ST8/01594.

References

1. Zhang W, He H, Xu C, Yu W, Li L (2019) Precipitates dissolution, phase transformation, and re-precipitation-induced hardness variation in 6082–T6 alloy during MIG welding and subsequent baking. JOM 71:2711–2720

2. Caceres CH, Davidson CJ, Griffiths JR, Newton CL (2002) Effects of solidification rate and ageing on the microstructure and mechanical properties of AZ91 alloy. Mater Sci Eng A 325(1–2):344–355
3. ASM metals handbook: alloy phase diagrams, vol 3. Materials Park, OH, ASM International, p 48 (1992)
4. McLean AA, Powell GLF, Brown IH, Linton VM (2013) Friction stir welding of magnesium alloy AZ31B to aluminium alloy 5083. Sci Technol Weld Join 8(6):461–464
5. Gerlich A, Su P, Noth TH (2005) Peak temperatures and microstructures in aluminium and magnesium alloy friction stir spot welds. Sci Technol Weld Join 10(6):647–652
6. Buffa G, Baffari D, Di Caro A, Fratini L (2015) Friction stir welding of dissimilar aluminium-magnesium joints: sheet mutual position effects. Sci Technol Weld Join 20(4):271–279
7. Sameer MD, Birru AK (2019) Mechanical and metallurgical properties of friction stir welded dissimilar joints of AZ91 magnesium alloy and AA 6082–T6 aluminium alloy. J Magnes Alloy 7:264–271
8. Park SK, Hong ST, Park JH, Park KY, Kwon YJ, Son HJ (2010) Effect of material locations on properties of friction stir welding joints of dissimilar aluminium alloys. Sci Technol Weld Join 15(4):331–336
9. Guo JF, Chen HC, Sun CN, Bi G, Sun Z, Wei J (2014) Friction stir welding of dissimilar materials between AA6061 and AA7075 Al alloys effects of process parameters. Mater Des 56(4):185–192
10. Reza-E-Rabby M, Tang W, Reynolds AP (2015) Effect of tool pin features on process response variables during friction stir welding of dissimilar aluminum alloys. Sci Technol Weld Join 20(5):425–432
11. Amancio-Filho ST, Sheikhi S, dos Santos JF, Bolfarini C (2008) Preliminary study on the microstructure and mechanical properties of dissimilar friction stir welds in aircraft aluminium alloys 2024-T351 and 6056-T4. J Mater Proc Tech 206(1–3):132–142
12. Hamilton C, Kopyściański M, Węglowska A, Pietras A, Dymek S (2019) Modeling, microstructure, and mechanical properties of dissimilar 2017A and 5083 aluminium alloys friction stir welds. J Eng Manu 233:553–564
13. Xu Y, Ke L, Ouyang S, Mao Y, Niu P (2021) Precipitation behavior of intermetallic compounds and their effect on mechanical properties of thick plate friction stir welded Al/Mg joint. J Manu Proc 64:1059–1069

A Simulation Study on Material Flow and Mixing Mechanism in Dissimilar Friction Stir Welding of AA6061 and AZ31 Alloys

Chengle Yang, Qingyu Shi, and Gaoqiang Chen

Abstract During dissimilar Friction Stir Welding (FSW), proper material flow and mixing processes are of critical importance for a sound joint, but its mechanism remains unclear. In this paper, a Computational Fluid Dynamics (CFD)-based simulation method was employed to study the material flow and mixing mechanism in dissimilar Friction Stir Welding of AA6061 and AZ31 alloys based on the numerical prediction. The predicted temperature field agrees well with the experiment. The temperature measured in the Mg on the advancing side (AS) is higher than that in Al on the retreating side (RS). During the welding process, an unsteady mixing process takes place near the welding tool. Two material mixing patterns during the welding process are predicted, which are shoulder-affected (SA) mixing and pin-affected (PA) mixing. The SA mixing is simple and on the macro scale which leads to the interlocking feature in the joint. By contrast, the PA mixing is more intense and results in a high-Mg-percentage mixture with Al-rich patches in the AS of the SZ.

Keywords Friction stir welding · Aluminium · Magnesium · Simulation

Introduction

Al alloys and Mg alloys have been widely applied in aerospace and manufacturing of vehicles because of their benefit in structure light-weighting, and hence, the demand for joining between them is expected [1]. However, the fusion welding of Al–Mg alloys has been reported to generate a thick brittle intermetallic compounds (IMCs) layer at the interface of the two alloys, resulting in the poor mechanical performance of the joint [2–4]. Friction Stir Welding (FSW) has been proven capable in dissimilar welding of Al and Mg alloys for its low heat input and mechanical effect which restrains the growth of the IMCs layer and generates an interface with a complex profile between the two alloys [5]. The preferable material mixture in the joint resulting from the severe material flow during the welding process is another

C. Yang · Q. Shi · G. Chen (✉)
Department of Mechanical Engineering, Tsinghua University, Beijing, China
e-mail: Cheng1@tsinghua.edu.cn

Y. Hovanski et al. (eds.), *Friction Stir Welding and Processing XII*,
The Minerals, Metals & Materials Series,
https://doi.org/10.1007/978-3-031-22661-8_21

primary factor in determining the joint strength [5–8]. The shape and internal structure of the mixing region strongly affected the fracture behavior and mechanical properties of the joint [9–11]. However, the process of how the mixing pattern forms remains unclear.

A great deal of previous research into dissimilar FSW of Al–Mg alloys has focused on the joint appearance [5, 8, 12, 13]. There is a consensus about the joint structure that a tortuous interface of Al–Mg exists in the middle of the stir zone (SZ), and a mixing region consisting of both materials is between the interface and the base material on the advancing side (AS). The marker material method was utilized to measure the flow variables including velocity and strain rate during the welding process [14]. Together, these studies indicate that the material in a different part of the joint has been through a different flowing process, resulting in different mixing patterns.

To establish a direct connection between material flow and joint appearance, efforts by simulation method have been made [15–17]. The welding temperature and the shape of the mixing region are well-predicted, which proves the validity of the Computational Fluid Dynamics (CFD) approach with the Volume of Fluid (VOF) model for tracing the material interface during dissimilar FSW of Al–Mg alloys. One potential problem in the current study is that the predicted material distribution is monotonic from the retreating side (RS) to the AS, and, so far, little attention has been paid to the structure of the complex mixing region in the SZ and its forming mechanism.

In this paper, a transient VOF model based on the CFD approach and a state-of-the-art frictional shear stress boundary condition at the tool-workpiece interface [18] were employed to investigate the material flow and mixing mechanism.

Method

The materials of the two workpieces are commercial AA6061-T6 Al alloy and AZ31B-H24 Mg alloy. The dimension of each workpiece is 150 × 50 × 3 mm^3 (length × width × thickness). The Mg workpiece was placed on the AS, and the Al workpiece was on the RS. The diameter of the tool shoulder is 13 mm. The tool pin is in a smooth frustum shape, with a root diameter of 4 mm, end diameter of 3.5 mm, and length of 2.8 mm. The FSW experiment was conducted at a tool rotation rate of 950 rpm and welding speed of 60 mm/min. The tool axis was aligned with the weld centerline. The plunge depth of the tool shoulder was 0.2 mm, and a tool tilt angle of 2.5° against the welding direction ensured the compact contact between the tool and the workpieces.

K-type thermocouples were embedded in the workpieces on both AS and RS at 8 mm from the weld centerline and 1.5 mm from the top surface to measure the temperature during the welding process.

The geometry and boundary definition of the CFD model is shown in Fig. 1a. The dimension of the geometry model is the same as the experiment. The volume

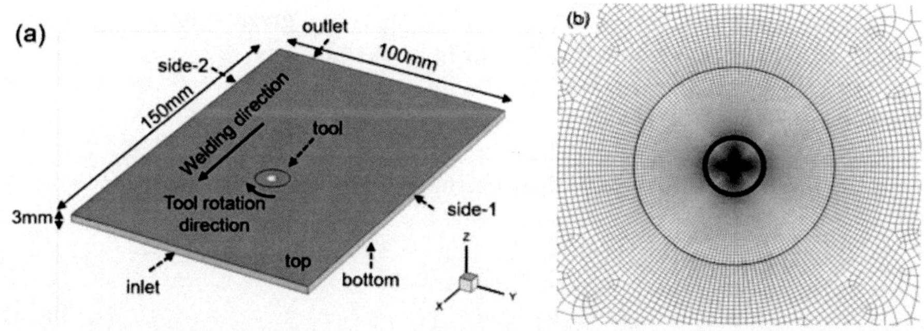

Fig. 1 Schematic of model. **a** Boundary definition and dimension of the model. **b** Mesh scheme near the welding tool (the red circle is the periphery of the tool shoulder)

of the tool pin was excluded from the fluid domain. The whole calculation domain has meshed into 510,024 hexagonal grids. The sizes of the grids near the tool are smaller than 0.1 mm to capture the intense material flow and mixing. A Cartesian coordinate system was established, whose X-axis aligns with the welding direction, Y-axis points to the AS, and the origin is at the intersection point of the tool axis and the top surface of the workpiece.

The Volume of Fraction (VOF) model was utilized. Transient governing equations including continuity equation, momentum equation, and energy equation are given by:

$$\frac{\partial \rho_{\text{mix}}}{\partial t} + \nabla \cdot (\rho_{\text{mix}} \vec{v}) = 0 \tag{1}$$

$$\frac{\partial (\rho_{\text{mix}} \vec{v})}{\partial t} + \nabla \cdot (\rho_{\text{mix}} \vec{v}\vec{v}) = -\nabla P + \nabla \cdot \left[\mu_{\text{mix}} \left(\nabla \vec{v} + \nabla \vec{v}^T\right)\right] \tag{2}$$

$$\frac{\partial (\rho_{\text{mix}} H_{\text{mix}})}{\partial t} + \nabla \cdot \left(\rho_{\text{mix}}\ \vec{v}\ H\right) = \nabla \cdot (k_{\text{mix}} \nabla T) + S_V \tag{3}$$

where t is the time, $\vec{v}$ is the velocity vector, and P is the pressure tensor.

According to assumptions in the VOF model, for any given material property ϕ_{mix} of the mixture, including density ρ_{mix}, viscosity μ_{mix}, heat conductivity k_{mix}, and enthalpy H_{mix}, the value is determined by the volume fraction of the two alloys, given by:

$$\phi_{\text{mix}} = \alpha_{\text{Mg}} \phi_{\text{Mg}} + \alpha_{\text{Al}} \phi_{\text{Al}} \tag{4}$$

where α_{Mg} and α_{Al} are volume fraction of each alloy, and ϕ_{Mg} and ϕ_{Al} are material properties, respectively.

The volume fraction of Al is determined by:

$$\frac{\partial (\alpha_{\text{Al}} \rho_{\text{Al}})}{\partial t} + \nabla \cdot (\alpha_{\text{Al}} \rho_{\text{Al}} \vec{v}_{\text{Al}}) = \dot{m}_{\text{Mg,Al}} - \dot{m}_{\text{Al,Mg}} \tag{5}$$

where $\dot{m}_{\mathrm{Mg,Al}}$ is the mass transfer from Mg to Al in a given volume in a unit time and vice versa. The volume fraction of Mg is given by:

$$\alpha_{\mathrm{Mg}} = 1 - \alpha_{\mathrm{Al}} \tag{6}$$

The viscosity of the material is determined by the viscoelastic theory given by [19]

$$\mu = \frac{\sigma}{3\dot{\varepsilon}} \tag{7}$$

where $\dot{\varepsilon}$ is the effective strain rate. σ is the flow stress of the material, given by [18]

$$\sigma = \sigma_p \sinh^{-1}\left[\left(\frac{\dot{\varepsilon}}{A}\exp\left(\frac{Q}{RT}\right)\right)^{\frac{1}{n}}\right] \tag{8}$$

where σ_p, A, and n are the material constants, Q is the activation energy of the material, and R is the ideal gas constant. An empirical softening of material is considered. The values of the material constants are given in Table 1. The constants of the AZ31B-H24 Mg alloy were determined by the method in [20] and from data in [21].

The thermal property of the material is the same as [17].

At the boundary of the tool-workpiece interface, the interfacial stress on the workpiece material provided by the welding tool is given by a state-of-the-art flow-mechanical coupled model in [18] as

$$\tau_f = M \cdot \vec{D} \cdot \beta \tag{9}$$

where M is the magnitude term. $\vec{D}$ is the direction term, and β is the pseudo-stick term, given by

$$\vec{D} = \frac{\vec{v}_{\mathrm{rel}}}{||\vec{v}_{\mathrm{rel}}||} \tag{10}$$

$$\beta = \tanh(\alpha \cdot ||\vec{v}_{\mathrm{rel}}||) \tag{11}$$

where $\vec{v}_{rel}$ is the relative velocity between the tool and workpiece at the interface, and α is a constant.

Table 1 Material constants

Alloy	σ_p, MPa	A	Q, J/mol	n	References
AA6061-T6	57.8	1.68×10^{21}	292,000	7.25	[20]
AZ31B-H24	200.0	2.96×10^{15}	16,530	5.42	[20, 21]

Other settings of the boundary condition and heat generation setting are similar to [22].

Results and Discussion

The predicted temperature field is shown in Fig. 2. A basin-shaped region of high temperature is in the vicinity of the tool-workpiece interface. The peak temperature is 305.3 °C. The temperature on the AS is higher than that on the RS. Along the welding direction, the material at the rear side of the tool has a higher temperature than the leading side.

The temperature measured in the experiment and prediction is compared in Fig. 3. The measured peak temperature is 231.9 °C on the AS and 209.7 °C on the RS. The predicted peak temperature is 225.7 °C and 210.4 °C, correspondingly. In FSW of a single material, the temperature at AS is slightly higher than at RS, and in the case of dissimilar FSW of Al–Mg alloys, Mg with lower heat conductivity on the AS than Al would expand this difference. The predicted peak temperature and high-temperature profile agree well with the experiment.

Compared to FSW of a single kind of Al [23] or Mg [24] alloy, whose welding temperature could exceed 450 °C, the temperature of Al–Mg dissimilar welding is much lower. In current studies of similar welding conditions, the peak welding temperature is about 350 °C near the pin side [12], and about 360 °C at the tool-workpiece interface [17], revealing the specificity of dissimilar FSW of Al–Mg alloys.

The predicted material flowing details near the tool is exhibited in Fig. 4. It could be found that, in Fig. 4a, the mixing of Al and Mg occurs on both the AS and the RS of the tool. The tool pin is closely surrounded by an Al-rich mixture. A piece of pure Mg is wrapped in the mixture on the RS. The viscosity in this Mg region is higher than the surrounding mixture with a higher proportion of Al, as could be found in Fig. 4b.

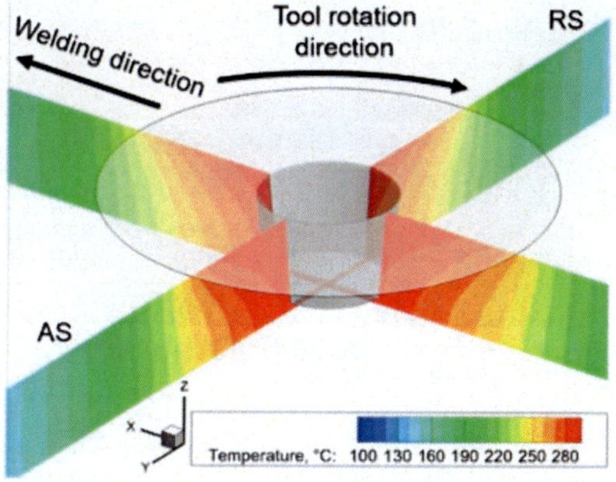

Fig. 2 Predicted temperature field

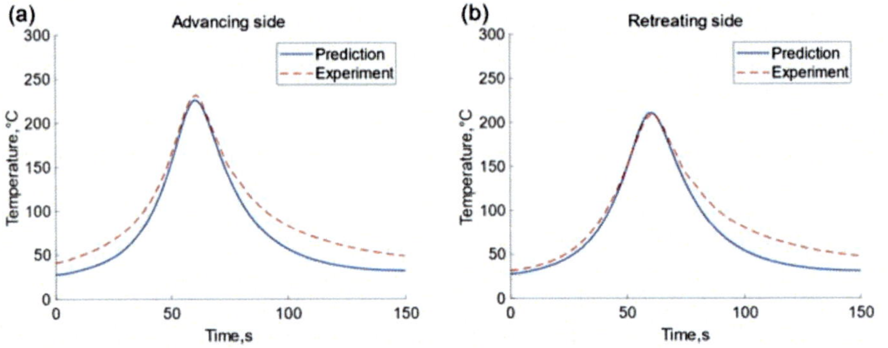

Fig. 3 Predicted and experiment measured welding temperature at 8 mm from the weld center and 1.5 mm from the upper surface of the workpiece on **a** the AS and **b** the RS

As shown in Fig. 4c, there is a thin layer of the high-velocity-flowing region next to the tool-workpiece interface near the shoulder-pin corner. The highest velocity in the whole flow field is predicted as 0.20 m/s in this region. The high-velocity layer is wider near the interface on the AS, but the motion is conducted further into the workpiece from the interface on the RS. Hence, as shown in Fig. 4d, a large strain rate of material occurs on both the RS and AS, while there is an additional high-strain-rate band at the edge of the mixing region on the RS, near the periphery of the tool shoulder. The highest strain rate is 2212.3 s^{-1} as predicted.

In the predicted flow field, the material is mixed and the mixture is formed in the joint, as shown in Fig. 5. Figure 5a presents the predicted material distribution in a transverse slice of the joint. There is a boundary in the middle of the joint and an Mg-rich mixture between the boundary and the Mg base material on the AS, which is following previous studies on the joint appearance [5, 8, 12, 13]. The result that the top surface of the joint is covered by a thin layer of Al is also consistent with recent experiment investigations [11]. At the upper part of the joint, material mixing is on

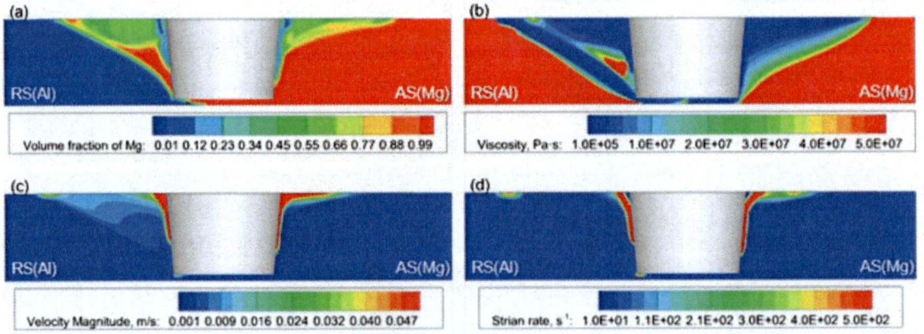

Fig. 4 Material flow field at X = 0 slice across the tool axis. **a** Material distribution by volume fraction of Mg. **b** Viscosity. **c** Velocity magnitude. **d** Strain rate

a macro scale, where a piece of Mg inserts into Al on the RS, and below that, a piece of Al inserts into Mg on the AS. This inter-inserting pattern forms the interlocking feature. This kind of mixing material is named shoulder-affected (SA) mixture. In the Mg-rich mixture on the AS of the SZ, there is a decrease in the Mg fraction, which indicates a rising in the Al fraction in the mixture and differs from previous simulation studies [15, 17]. This mixture of a high percentage of Mg with a small Al-rich patch in it is named as pin-affected (PA) mixture. The mixing mechanism of these two kinds of mixture is thus named SA mixing and PA mixing, respectively.

Figure 5b exhibits a longitudinal view of the material distribution along the weld. It could be found that the material flow during the welding process is unsteady since the contour surface is in a fluctuating shape in the joint. The contour ridges of the SA mixture, as indicated, are of different lengths, and the Al-rich regions in the PA mixture, shown as bands or fragments with closed contour surface, are discontinuous and of different sizes and shapes. In Fig. 5c, the material mixing process is presented. On the RS of the pin, the PA mixing occurs between the pure Al connecting with base metal and mixture with a large percentage of Mg, which forms a semi-circle-shape sharp boundary. This boundary is distributed to the mixture by the transverse motion of the shoulder on the rear side of the pin, becoming part of the SA mixture. Around the pin, the PA mixing occurs between the circular Al-rich material next to the pin and the surrounded Mg. On the AS of the pin, the mixture with a higher proportion of Al deposits forms the PA mixture in the joint.

To demonstrate the material flow behavior of the mixing process, streamlines are extracted from the flow field as shown in Fig. 6. In Fig. 6a, the SA mixing is along large radius semi-circle paths, which are near the shoulder and bypass the pin, indicating a simple and smooth flowing pattern of SA mixing. However, for the PA mixing in Fig. 6b, the selected streamline revolves around the tool pin for multiple

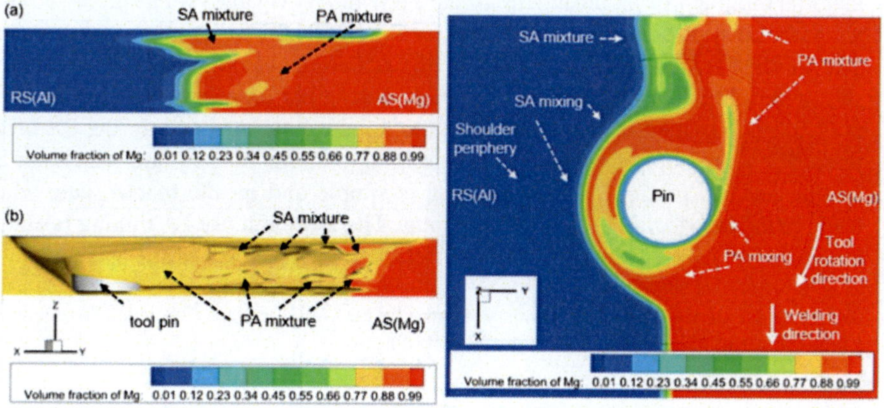

Fig. 5 Material mixing during the welding process and distribution in the joint. **a** Predicted material distribution in the joint at X = −10 mm slice. **b** A 3D oblique view of material distribution from the tool to the slice in **a**, with the contour surface at a volume fraction of Mg being 0.8. **c** Material mixing and mixture formation near the tool pin, at the horizontal slice plane of Z = −1 mm

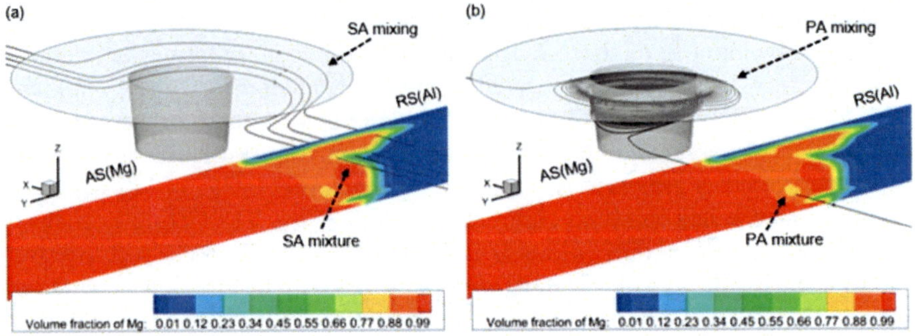

Fig. 6 Material flow pattern of different mixing mode and their corresponding depositing position at the slice of Fig. 5a. **a** SA mixing. **b** PA mixing

circles in a smaller radius. Along the streamline, the radius decreases and goes down in the negative Z-direction. The multi-circular part of the streamline is in the high-velocity region near the tool-workpiece interface, as shown in Fig. 4c. This reveals the PA mixing is more intense than the SA mixing. It is worth noting that, although the streamlines might not be the same as the real material flowing trace because of the unsteadiness of the flow field, they could still represent the feature of the two mixing mechanisms and the difference between them.

Conclusion

A CFD model based on VOF formulation has been established to investigate the material mixing process during dissimilar FSW of Al–Mg alloys. The predicted peak temperature and temperature profile agree well with the experiment measurement. The welding temperature on the AS is higher than that on the RS. The simulation shows that the material mixing is unsteady during the FSW process, resulting in a fluctuant component distribution along the longitudinal direction in the SZ of the joint. The simulation demonstrates two mixing patterns in the joint, which are SA mixing and PA mixing. The SA mixing is simple and on the macro scale which leads to the interlocking feature in the joint. The more intense PA mixing results in a high-Mg-percentage mixture with Al-rich patches in the AS of the SZ.

Acknowledgements The research was supported by the National Natural Science Foundation of China (grant no. 52175334 and grant no. 52035005).

References

1. Shankar S, Mehta KP, Chattopadhyaya S, Vilaça P (2022) Dissimilar friction stir welding of Al to non-Al metallic materials: an overview. Mater Chem Phys 288:126371
2. Borrisutthekul R, Miyashita Y, Mutoh Y (2005) Dissimilar material laser welding between magnesium alloy AZ31B and aluminium alloy A5052-O. Sci Technol Adv Mater 6:199–204
3. Liu L, Ren DX, Liu F (2014) A review of dissimilar welding techniques for magnesium alloys to aluminum alloys. Materials 7:3735–3757
4. Baqer YM, Ramesh S, Yusof F, Manladan SM (2018) Challenges and advances in laser welding of dissimilar light alloys: Al/Mg, Al/Ti, and Mg/Ti alloys. Int J Adv Manufact Technol 95:4353–4369
5. Shah LH, Othman NH, Gerlich A (2017) Review of research progress on aluminium–magnesium dissimilar friction stir welding. Sci Technol Weld Joining 23:256–270
6. Bruck HA (2004) Using geometric complexity to enhance the interfacial strength of heterogeneous structures fabricated in a multi-stage, multi-piece molding process 44:261–271
7. Wu A, Song Z, Nakata K, Liao J, Zhou L (2015) Interface and properties of the friction stir welded joints of titanium alloy Ti6Al4V with aluminum alloy 6061. Mater Des 71:85–92
8. Zhao Y, Lu Z, Yan K, Huang L (2015) Microstructural characterizations and mechanical properties in underwater friction stir welding of aluminum and magnesium dissimilar alloys. Mater Des (1980–2015) 65:675–681
9. Dorbane A, Mansoor B, Ayoub G, Shunmugasamy VC, Imad A (2016) Mechanical, microstructural and fracture properties of dissimilar welds produced by friction stir welding of AZ31B and Al6061. Mater Sci Eng, A 651:720–733
10. Fu B, Qin G, Li F, Meng X, Zhang J, Wu C (2015) Friction stir welding process of dissimilar metals of 6061–T6 aluminium alloy to AZ31B magnesium alloy. J Mater Process Technol 218:38–47
11. Abdollahzadeh A, Shokuhfar A, Cabrera JM, Zhilyaev AP, Omidvar H (2019) In-situ nanocomposite in friction stir welding of 6061–T6 aluminum alloy to AZ31 magnesium alloy. J Mater Process Technol 263:296–307
12. Lv X, Wu C, Sun Z (2022) Effects of ultrasonic vibration on material flow and thermal cycles in friction stir welding of dissimilar Al/Mg alloys. Metall Mater Trans A 53:1572–1584
13. Xiaoqing J, Yongyong L, Tao Y, Shujun C, Lei W, Wang J (2022) Enhanced mechanical properties of dissimilar Al and Mg alloys fabricated by pulse current assisted friction stir welding. J Manuf Process 76:123–137
14. Wu C, Wang T, Su H (2022) Material flow velocity, strain and strain rate in ultrasonic vibration enhanced friction stir welding of dissimilar Al/Mg alloys. J Manuf Process 75:13–22
15. Yang CL, Wu CS, Lv XQ (2018) Numerical analysis of mass transfer and material mixing in friction stir welding of aluminum/magnesium alloys. J Manuf Process 32:380–394
16. Yang C, Wu C, Shi L (2020) Modeling the dissimilar material flow and mixing in friction stir welding of aluminum to magnesium alloys. J Alloy Compd 843
17. Jiang T, Wu C, Shi L (2022) Effects of tool pin thread on temperature field and material mixing in friction stir welding of dissimilar Al/Mg alloys. J Manuf Process 74:112–122
18. Chen G, Feng Z, Zhu Y, Shi Q (2016) An alternative frictional boundary condition for computational fluid dynamics simulation of friction stir welding. J Mater Eng Perform 25:4016–4023
19. Colegrove PA, Shercliff HR (2005) 3-Dimensional CFD modelling of flow round a threaded friction stir welding tool profile. J Mater Process Technol 169:320–327
20. Ding SS, Shi QY, Chen GQ (2021) Flow stress of 6061 aluminium alloy at typical temperatures during friction stir welding based on hot compression tests. Metals 11
21. Prasad YVRK, Rao KP, Sasidhara S. Hot working guide—a compendium of processing maps, 2nd edn. ASM International
22. Yang C, Dai Q, Shi Q, Wu C, Zhang H, Chen G (2022) Flow-coupled thermo-mechanical analysis of frictional behaviors at the tool-workpiece interface during friction stir welding. J Manuf Process 79:394–404

23. Khan NZ, Bajaj D, Siddiquee AN, Khan ZA, Abidi MH, Umer U, Alkhalefah H (2019) Investigation on effect of strain rate and heat generation on traverse force in FSW of dissimilar aerospace grade aluminium alloys. Mater (Basel) 12
24. Hasan AF (2019) CFD modelling of friction stir welding (FSW) process of AZ31 magnesium alloy using volume of fluid method. J Market Res 8:1819–1827

Mechanism of Joint Formation in Dissimilar Friction Stir Welding of Aluminum to Steel

Amlan Kar, Todd Curtis, Bharat K. Jasthi, Wade Lein, Zackery McClelland, and Grant Crawford

Abstract This paper highlights the influence of load-controlled experiments on heat input and corresponding improvement in mechanical properties of dissimilar friction stir lap welding of aluminum 6061 alloy (Al-6061) and mild steel (MS) with different thicknesses. Dissimilar welds are produced at different loads and heat input conditions. The microstructure and chemical composition of the weld interface were characterized to determine the deformation behaviour, quality of interfacial layer formation, and joining mechanisms at the aluminum-to-steel interface. An improvement (more than 65% joint efficiency) in shear tensile properties was reported under optimized process conditions. Mechanisms associated with improved tensile properties have been identified. Mechanical mixing, the formation of intercalated structures, and the evolution of diffusion layers are considered the mechanisms of dissimilar joint formation.

Keywords Dissimilar friction stir weld · Load control experiments · Microstructure · Tensile properties · Mechanical mixing · Mechanisms · Intercalated structures

Introduction

Dissimilar welding of aluminum (Al) to steel (Fe) provides the potential to lightweight structural components while maintaining the excellent strength and corrosion resistance of the parent materials. Therefore, welding of Al to Fe is vital

A. Kar · T. Curtis · B. K. Jasthi · G. Crawford (✉)
Arbegast Materials Processing and Joining Laboratory (AMP), South Dakota School of Mines & Technology, Rapid City, SD 57701, USA
e-mail: Grant.Crawford@sdsmt.edu

W. Lein
U.S. Army Corps of Engineers Cold Regions Research and Engineering Laboratory, Vicksburg, MS, USA

Z. McClelland
U.S. Army Engineer Research and Development Center, Vicksburg, MS, USA

Y. Hovanski et al. (eds.), *Friction Stir Welding and Processing XII*,
The Minerals, Metals & Materials Series,
https://doi.org/10.1007/978-3-031-22661-8_22

for a wide range of industrial applications (e.g., aerospace, automotive, petrochemical, and energy) [1]. Dissimilar welded structures offer design flexibility, which may result in improved location-specific strength, significant weight reduction, and, in many cases, reduced production cost. However, welding of Al to Fe using conventional fusion welding techniques has historically been problematic and cost-prohibitive. This is mainly due to the large difference in their physical properties, thermal properties, and formation of brittle intermetallic compounds (IMC). The IMCs at the weld interface cause crack formation and propagation when subjected to external mechanical loading, leading to inferior mechanical properties of the dissimilar metal joints.

Solid-state welding has been used in recent years to join different similar and dissimilar materials under an applied load at a temperature below the melting point of all contacting materials. Friction stir welding (FSW) is a solid-state welding technique that has been extensively used for dissimilar metal welding [2, 3]. It is recognized as green technology with numerous advantages, including no need for shielding gases, low tenacity of forming solidification defects, and low heat input. Despite numerous benefits, dissimilar FSW of Al to Fe is frequently associated with low mechanical properties due to the formation of interfacial defects, inhomogeneous elemental mixing at the interface leading to crack formation, and evolution of brittle IMCs [2].

Numerous studies have evaluated defect formation and characterized IMC development in dissimilar Al to Fe FSWs [4–7]. However, little attention has been paid to understanding the influence of interface characteristics on the mechanical properties of dissimilar Al to Fe FSWs. The quality of the interface determines defect formation and the type of IMCs that develop. During tensile testing, the composition and local microstructure present in each layer across the weld interface determine the location of crack initiation. A brittle interface and weaker zone across the weld may result in the formation of cracks and premature failure of the joints leading to inferior mechanical properties. On the other hand, the chemical composition and mechanical mixing across the interface can encourage the formation of localized diffusion layers, and associated IMC formation, depending on the thermomechanical condition of the system [5]. Thermomechanical conditions, including heat input and severity of deformation at the interface in FSW, can be controlled by processing parameters, design of welding interfaces [1], normal load, feedback torques, and a number of welding passes, etc. One of the difficulties in friction stir lap welding (FSLW) of soft to hard materials, such as welding of Al to Fe, is tool wear that leads to variation in deformation and process conditions from one end to the other end of the weld. Most of the past research work avoided severe interaction of rotating tool with Fe alloy to avoid tool wear or provided little penetration into Fe plate in FSLW of Al to Fe [6, 8]. Refractory metal pin tools were extensively used to protect the tool materials from overheating and wearing out during welding [6, 9]. Instead of over-plunging, multipass welding could be employed to protect the tool, increase the bonding area, promote mechanical joining, and improve mechanical properties [7, 10].

In the current investigation, multipass FSLW of aluminum 6061 (Al-6061) to mild steel (MS) was performed to elucidate the influence of heat input and interface characteristics on the mechanical properties of the dissimilar Al-6061/MS weld. Therefore, multipass FSLW up to three passes with 100% overlapped were conducted to vary the input conditions and microstructural evolution at the weld interface. The effect of weld microstructure and chemical composition on mechanical properties was investigated to determine the mechanism of joint formation and elucidate ideal joint quality for superior mechanical properties.

Experimental Procedure

Aluminum alloy AA6061-T6 (Al-6061) and mild steel (MS) sheet has been lap welded up to three passes using an i-STIR-10 FSW system. A schematic drawing of the process is shown in Fig. 1. The Al and Fe sheet thickness was 3.175 mm and 12.7 mm, respectively. Before FSLW, the surface contamination of Al and Fe plates was polished by abrasive paper and cleaned by acetone. The Al sheet was lapped to the Fe sheet at the advancing side (AS) for lap welding. A conventional frustum-shaped pin tool made of H13 tool steel with 25.4 mm shoulder diameter and convex feature was used for the welding. The length and diameter of the tool pin were 3.0 mm, and varied from 6 mm at the root and 4 mm at the tip. Load control FSW was performed at 400 rpm (revolution/min) tool rotation speed, 127 mm/min welding speed, and a constant value of tool tilt angle at 2.5°.

The welded sample was sectioned using a waterjet cutting machine for metallographic characterization and shear tensile testing, as schematically shown in Fig. 1. The metallographic samples with different welding passes were initially polished with the help of silica sandpapers of different grit size of 240, 400, 600, and 1200. Final polishing was conducted by diamond cloth polishing. The joint interface of the weld samples was examined with the help of a dual beam field emission scanning

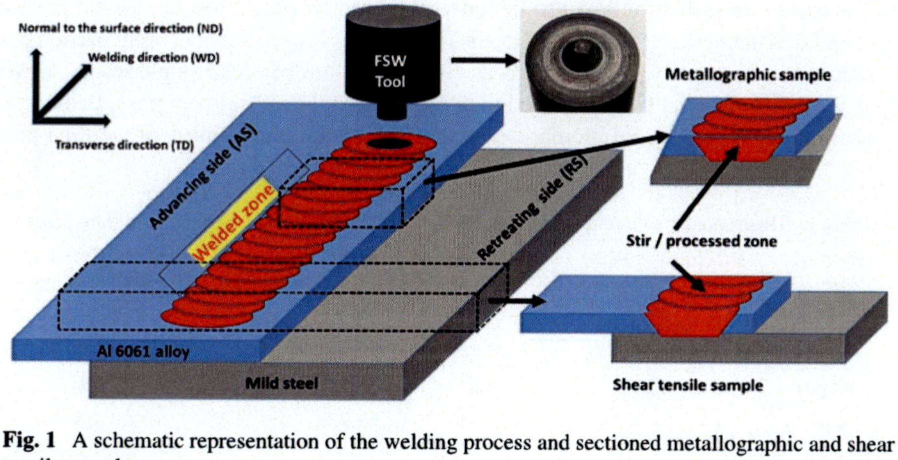

Fig. 1 A schematic representation of the welding process and sectioned metallographic and shear tensile samples

electron microscope (Helios 5 CX, Thermo Scientific, Massachusetts, USA) integrated with Oxford energy dispersive spectroscopy (EDS) detectors and software. The dimensions of the shear tensile sample with different welding passes and heat input were 150 mm in length and 24 mm in width along a line perpendicular to the welding direction. The tests were performed using a servo-hydraulic uniaxial testing system (Inc., Model 313 (Shakopee, MN, USA)) at room temperature using a displacement rate of 1.27 mm/min to evaluate the shear strength of the welds.

Results and Discussion

Heat Input Calculation

The main goal of FSLW is to generate dissimilar lap welds between Al-6061 and MS with improved bond strengths by generating thermal energy due to friction. The generated thermal energy acts as heat input to the welding system, which soften the weld plates and helps in stirring the material around the FSW tool. Therefore, heat input is an essential factor in process optimization, microstructure evolution, and final mechanical properties of the welds. In FSW, heat input is calculated using Eq. (1) [11].

$$\text{Heat input} = \frac{\text{Power}}{\text{Welding speed}} = \frac{\text{Tool rotation speed} \times \text{Torque}}{\text{Welding speed}} \tag{1}$$

In this equation, tool rotation speed (TRS) and welding speed (WS) are extrinsic parameters whereas torque is an intrinsic parameter coming from the deformation of materials around the tool. The theoretical model described here is extensively used in literature and correlates greatly with experimental results. Considering the processing condition and experimentally obtained feedback torque, heat input was calculated using Eq. 1, as represented in Table 1. It is found that heat input in each pass reduces continuously with the increase in the number of welding passes. However, as the same material was subjected to processing in the subsequent pass, cumulative heat input to the material increases with the increase in the number of welding pass.

Table 1 Theoretically calculated heat input values for the welds with different welding passes

Weld ID	TRS (rpm)	WS (mm/min)	Forge load (kN)	Torque (Nm)	Heat input (kJ/m)	Cumulated heat input (kJ/m)
1st pass	400	127	23.13	90.83	286	286
2nd pass			24.91	88.67	279	565
3rd pass			24.46	83.92	264	829

Microstructural Characterization

The interface microstructures of the friction stir weld between Al 6061 alloy (Al-6061) and mild steel (MS) at different passes are shown in Fig. 2. The interfacial morphology of the welds varied with the number of passes and heat input conditions. An increase in the number of welding passes promoted additional heat input to the weld, deformation at the weld interface, and fragmentation of coarse particles developed in the earlier pass [10]. However, macroscale defects were not observed in any of the welds. After the first pass, the dissimilar Al-6061/MS FSW exhibited a straight interface, without a diffusion layer, and severe deformation occurred on the Fe side of the interface (Fig. 2a). The weld nugget exhibited several particles of different sizes (indicated by arrows in Fig. 2b), which were distributed in the Al matrix. These particles were seen in all welds with different processing conditions and heat inputs. After the second pass (Fig. 2b), an intercalated zone (ICZ) could be seen on the Fe side of the weld. The ICZ is the zone combining fragmented Fe and diffused Al leading to comparatively stronger joining [2]. This is the potential zone for defect formation if the processing parameters and flow behaviours are not properly optimized. The Al-6061/MS FSW, after the third pass (Fig. 2c), contained an additional zone, named mechanically mixed zone (MMZ). The MMZ is mainly characterized by a composite mixture of diffused particles, Fe particles, and Al matrix. It is considered the strongest joint quality in dissimilar welds [2].

Chemical Analysis of Interface

A mechanically mixed layer (MML) formed at the weld interface due to the inclusion of Fe particles and diffused particles, in the Al matrix, as described in Fig. 2c. An EDS line scan across an MML, formed at the weld interface after the third pass, showed a continuous composition variation from the Al to Fe side of the weld interface (Fig. 3a). Deviation from a constant elemental composition indicated the absence of IMCs. A shift from the continuous variation in composition illustrated the distribution of fine particles and diffused particles in the MML.

On the other hand, ICZ exhibits a more abrupt variation in elemental concentration (Fig. 3b). A constant elemental composition at certain regions of the ICZ indicates elemental diffusion and the possible formation of IMC of Al–Fe depending on thermomechanical conditions and Gibbs free energy [12, 13]. Severe deformation and forged diffusion of Al in the fragmented Fe flakes at the weld interface promote the formation of ICZ. Therefore, it can be said that the ICZ is more prone to the formation of an extended diffusion layer in comparison to the MMZ. Interestingly, these layer thicknesses reach the critical thickness of 10 μm to obtain a better mechanical property due to alternate layers of Al and Fe [14]. Therefore, the presence of ICZ generally improves the tensile properties due to the clutching of Al and Fe layers.

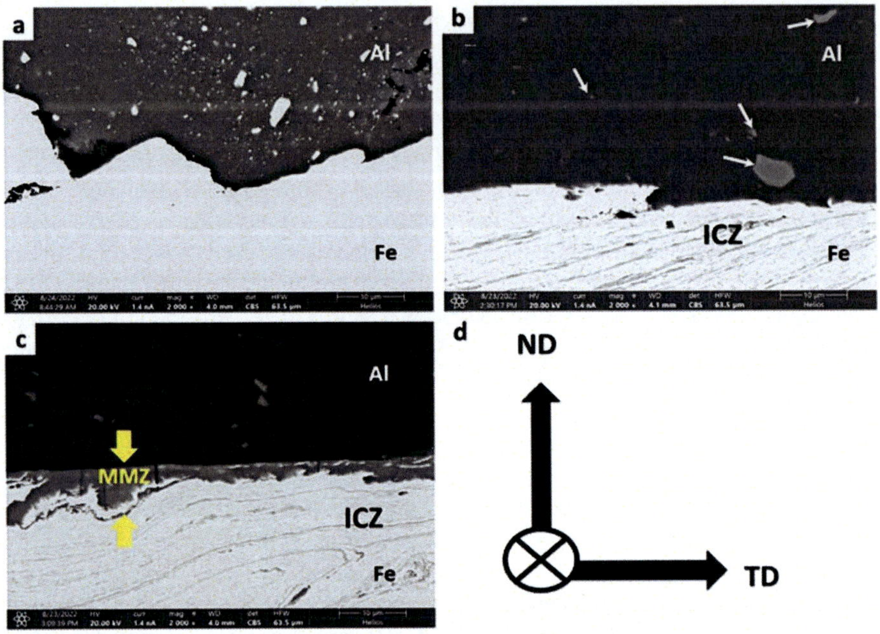

Fig. 2 Macroscopic cross-sectional SEM (backscatter) image of dissimilar Al-6061/MS FSW after **a** 1st pass, **b** 2nd pass, and **c** 3rd pass showing the difference in weld interfacial morphologies at **d** the ND-TD orientation of the welded plates (Fig. 1)

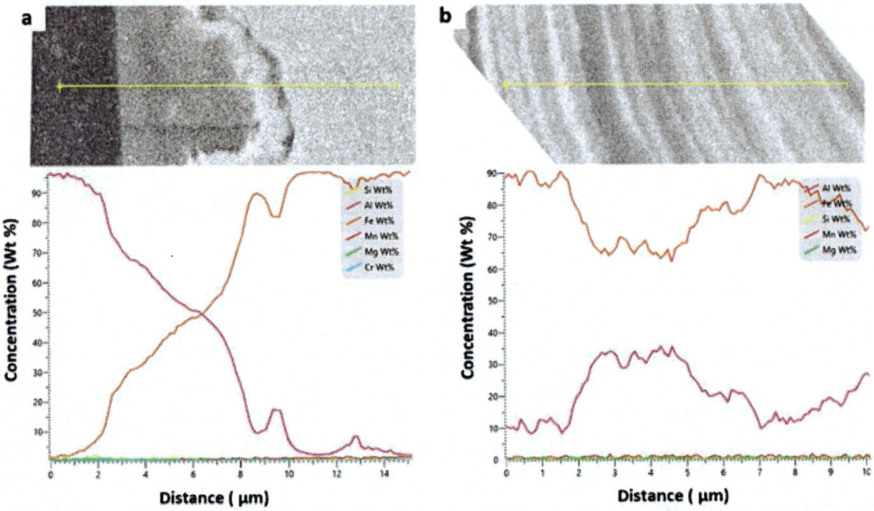

Fig. 3 EDS line scan analysis across the **a** mechanically mixed zone (MMZ) including IMCs and **b** intercalated zone (ICZ) showing elemental inhomogeneity across the different dissimilar weld interfacial zones

Mechanical Properties

The effect of welding passes and torque-based heat input on the shear tensile property can be seen in Fig. 4a. The weld after the first pass and after the second pass exhibited the lowest and highest shear tensile strength. The weld after the second pass exhibited joint strength of 196 MPa, which was 65% joint efficiency, compared with the base Al6061 having a tensile strength of 303 MPa. The same material also shows the lowest error (±3 MPa) in the tensile property. The low error bar indicates consistency in the tensile property. Hence, the material after the second pass is considered the best quality weld in the present investigation. The samples after shear tensile tests are shown in Fig. 4b. The weld after the first and third pass, which showed inferior welds, promoted crack formation and propagation in the RS of the Al plate within the processed zone. The welding after third pass showed joint strength of 157 MPa and 51% joint efficiency. The fracture location for the weld with second pass was outside the welding zone. Interestingly, no fracture at the Al-6061/MS joint was noticed, and that indicated the formation of a high-quality weld after each welding pass. Information from the fracture samples pointed to superior quality weld. The microstructure evolution in Al decided the fraction location, which was dictated by the cumulative heat input and thermomechanical deformation condition after each pass.

After the second pass, ICZ was formed in the weld interface, making the joint stronger. A similar weld interface with more deformation and diffusion layer formation could be seen after the third pass. The evolution of a thin diffusion layer along the weld interface made the joint weak. Further, the highest cumulative heat input in the weld after the second pass could lead to grain growth, which helped in reducing the weld's tensile properties due to the Hall–Petch strengthening mechanism [15]. The improvement in joint strength after the second pass, therefore, has been attributed to

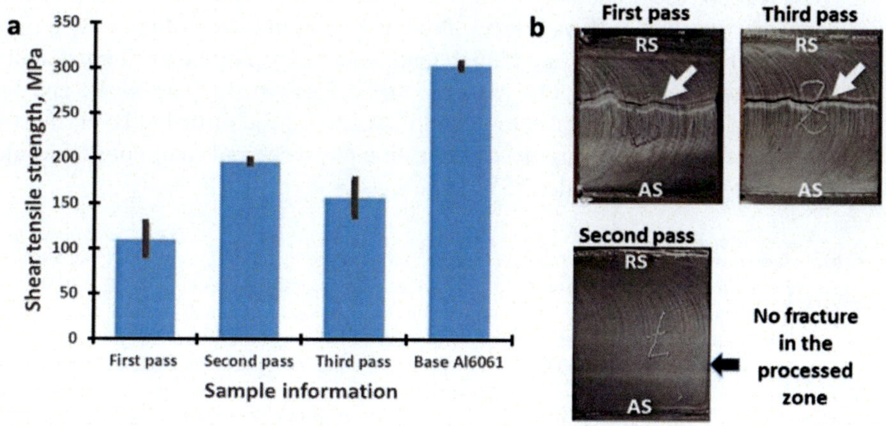

Fig. 4 Shear tensile test results showing **a** comparative tensile properties and **b** corresponding fracture samples with three different passes

the formation of ICZ at the joint interface, reduction in IMC, lesser grain growth, and appearance of fracture outside of the weld nugget (specifically in the heat-affected zone of the weld), etc.

Conclusion

Load-controlled friction stir lap welding with multiple welding passes and heat input conditions was performed to correlate process conditions and interface characteristics with the mechanical properties of the welds. Weld interfaces were characterized to understand the quality of the weld and chemical composition across the welds. The following conclusions can be drawn from the investigation.

1. Load-controlled, friction stir lap welding of Al 6061 to mild steel performed in multiple passes resulted in a variation in feedback torque and heat input and ultimately modified the thermomechanical process history imposed on the joint.
2. With increasing welding passes, during which deformation and heat input condition varies, led to forming three different layers, such as intercalated zone (ICZ), mechanically mixed zone (MMZ) and extended diffusion layer. The ICZ formed in the weld after the second pass. The weld with the third pass exhibited MMZ, ICZ, and alternating diffusion layers across the lap weld interface.
3. The diffusion layer formed in the ICZ after the second pass was thinner than the diffusion formed adjacent to the MML following the third welding pass. The presence of alternating diffused layers of parent materials in the ICZ made the joint stronger, and hence, the shear tensile sample fractured outside the welding zone.
4. The weld after the second pass exhibited 65% joint efficiency due to the formation of a thin diffusion layer and ICZ. The presence of MMZ after the third welding pass reduced the probability of interfacial defect formation; however, the development of a thick diffusion layer along the weld interface (due to higher heat input and extended exposure to high temperatures) weakened the joint, resulting in 51% joint efficiency. The presence and relative size of the MMZ and ICZ layers, in addition to the location and thickness of the diffusion layer, must be engineered through an appropriate selection of processing conditions to develop a strong Al to Fe dissimilar FSW joint.

Acknowledgements Funding for this research was supported by the US Army Engineer Research and Development Center. The use of trade, product, or firm names in this document is for descriptive purposes only and does not imply endorsement by the U.S. Government. The tests described and the resulting data presented herein, unless otherwise noted, are based upon work supported by the US Army ERDC under PE 0603119A, "Ground Advanced Technology", Task "Materials and Manufacturing Technology for Cold Environments". Permission was granted by the Director, Cold Regions Research and Engineering Laboratory to publish this information. The findings of this paper are not to be construed as an official Department of the Army position unless so designated by other authorized documents.

References

1. Ross K, Reza-E-Rabby M, McDonnell M, Whalen SA (2019) Advances in dissimilar metals joining through temperature control of friction stir welding. MRS Bulletin 44(8):613–618
2. Kar A, Vicharapu B, Morisada Y, Fujii H (2020) Elucidation of interfacial microstructure and properties in friction stir lap welding of aluminum alloy and mild steel. Mater Charact 168:110572
3. Kar A, Kailas SV, Suwas S (2019) Multi-length scale characterization of microstructure evolution and its consequence on mechanical properties in dissimilar friction stir welding of titanium to aluminum. Metall Mater Trans A 50:5153–5173
4. Wang X, Lados DA (2022) Optimization of aluminum-to-steel friction stir lap welding for the fabrication of high-integrity structural components. J Adv Joining Process 5:100114
5. Zhou L, Yu M, Liu B, Zhang Z, Liu S, Song X, Zhao H (2020) Microstructure and mechanical properties of Al/steel dissimilar welds fabricated by friction surfacing assisted friction stir lap welding. J Market Res 9(1):212–221
6. Shamsujjoha M, Jasthi BK, West M, Widener C (2015) Friction stir lap welding of aluminum to steel using refractory metal pin tools. J Eng Mater Technol 137(2). https://doi.org/10.1115/1.4029571
7. Leitao C, Arruti E, Aldanondo E, Rodrigues DM (2016) Aluminum-steel lap joining by multipass friction stir welding. Mater Des 106:153–160
8. Jana S, Hovanski Y, Grant GJ (2010) Friction stir lap welding of magnesium alloy to steel: a preliminary investigation. Metall Mater Trans A 41(12):3173–3182
9. Shamsujjoha M, Jasthi BK, West M, Widener C (2016) Microstructure and mechanical properties of FSW lap joint between pure copper and 1018 mild steel using refractory metal pin tools. In: Mishra R, Mahoney MW, Sato Y, Hovanski Y, Verma R (eds) Friction stir welding and processing VII. Springer International Publishing, Cham, pp 151–160
10. Kar A, Kailas SV, Suwas S (2018) Two-pass friction stir welding of aluminum alloy to titanium alloy: A simultaneous improvement in mechanical properties. Mater Sci Eng, A 733:199–210
11. Kumar S, Roy BS (2022) Force and torque perspective and their effects on the mechanical and microstructural characteristics of dissimilar friction stir welded acrylonitrile butadiene styrene and polycarbonate plates. Trans Indian Inst Met 75(3):843–854
12. Kar A, Kailas SV, Suwas S (2020) Formation sequence of intermetallics and kinetics of reaction layer growth during solid state reaction between titanium and aluminum. Materialia 11:100702
13. Kar A, Kailas SV, Suwas S (2019) Effect of mechanical mixing in dissimilar friction stir welding of aluminum to titanium with zinc interlayer. Trans Indian Inst Met 72(6):1533–1536
14. Mahto RP, Kumar R, Pal SK, Panda SK (2018) A comprehensive study on force, temperature, mechanical properties and micro-structural characterizations in friction stir lap welding of dissimilar materials (AA6061-T6 & AISI304). J Manuf Process 31:624–639
15. Kar A, Kailas SV, Suwas S (2018) Microstructural modification and high-temperature grain stability of aluminum in an aluminum-titanium friction stir weld with zinc interlayer. JOM 71:444–451

Part VIII
Derivative Technologies

Submerged Bobbin Tool (SBT) Tunneling Technology

Dwight A. Burford, Maurizio Manzo, Hector Siller, Supreeth Gaddam, Anurag Gumaste, James Koonce III, Aleandro Saez, and Rajiv S. Mishra

Abstract Submerged bobbin tool (SBT) tunneling is a new friction stir processing (FSP) technique for making integral channels within malleable materials. Like a conventional bobbin tool (BT) for friction stir welding (FSW), an SBT toolset has two opposing shoulders spaced apart along the bobbin or probe section of the tool. Unlike a conventional BT, an SBT is used to form integral subsurface channels by passing the shoulder at the distal end of the probe through the workpiece during processing. Example uses of internal pathways are found in heat exchangers, cooling plates, and vacuum tools. Advance uses may include lessening weight and modifying the stiffness of structural components. A preliminary evaluation in AA6061-T6511 plates shows this special form of FSP has low process forces and is therefore capable of being deployed on CNC (computer numerical control) machining centers and friction stir-capable industrial robots as well as purpose-built FSP machines. Consequently, SBT tunneling holds potential use in a wide range of applications requiring curvilinear internal pathways for wiring, gases, and fluids, as well as internal spaces for the placement of powders and solid materials like composites.

Keywords Friction stir processing · FSP · Friction stir channeling · FSC · Bobbin tools · Thermomechanical processing · Localized extruding · Heat exchangers · Cooling plates · Channels · Internal pathways

---

D. A. Burford (✉) · S. Gaddam · A. Gumaste · R. S. Mishra
Materials Science & Engineering, College of Engineering, University of North Texas, Discovery Park, Denton, TX 76207, USA
e-mail: dwight.burford@unt.edu

M. Manzo · H. Siller · J. Koonce III · A. Saez
Mechanical Engineering, College of Engineering, University of North Texas, Discovery Park, Denton, TX 76207, USA

D. A. Burford · M. Manzo · H. Siller · R. S. Mishra
Advanced Materials & Manufacturing Processes Institute (AMMPI), University of North Texas, Discovery Park, Denton, TX 76207, USA

Y. Hovanski et al. (eds.), *Friction Stir Welding and Processing XII*,
The Minerals, Metals & Materials Series,
https://doi.org/10.1007/978-3-031-22661-8_23

Introduction

Producing internal passageways in heat exchangers and similar hardware using friction stir processing (FSP) was introduced by Mishra in 2002 [1–3]. Existing methods available at the time for forming enclosed channels in metals involved fabricating matching sets of parts by stamping, extruding, etching, machining, and other manufacturing methods [4–6]. Once completed, the matching sets were then joined with fasteners, adhesives, laser welding, etc., to enclose the channel(s). Tubing attached mechanically or by brazing to metal plates has also been used in the production of heat management systems, such as in loop heat pipes [7].

This new use for FSP introduced by Mishra, known as friction stir channeling (FSC), was characterized by Balasubramanian et al. [8]. Using a commercially available CFD software package, the authors evaluated FSC channels having different hydraulic diameters. Samples were tested for both pressure drop and heat transfer along the lengths of the channels. This work was complemented by research characterizing the shape and surface roughness of FSC channels as a function of processing parameters [9]. The cross-sectional areas of the channels tested in this research were found to increase with traversing speed and a reduction in tool rotational rate.

Using an infrared (IR) camera, Balasubramanian et al. also studied peak temperatures in the process zone as a function of specific energy input into AA6061-T6 coupons during FSC [10]. From this research, a mechanistic model for spindle torque versus specific energy was developed and tested as a function of (1) the rotational rate of the FSC tool, (2) its traversing speed, and (3) its plunge depth within the workpiece. The model was based on a least-squares analysis, and test results showed a positive linear correlation between peak temperature and specific energy during FSC processing. It was also observed that the cross-sectional area of the channels decreased with increasing specific energy input.

FSC process forces in AA6061-T6 were also studied by Balasubramanian et al. [11]. Using a high-speed data acquisition system, the net resultant force acting on the tool was recorded and then plotted as a function of process parameters. The authors observed a correlation between material flow and the formation of channels in the stir zone based on the magnitude and direction of net forces acting on the tool during FSC. Internal channels were successfully produced when the resultant process force acting on the toolset was observed to be between the retreating side and the trailing edge of the tool probe. In contrast, only partial channels were produced when the resultant force was oriented between the trailing edge and the advancing side.

Pandya et al. used X-ray computed tomography and optical microscopy to study the morphology of channels produced by FSC [12]. In one instance, a broken probe embedded within the workpiece provided a unique opportunity to view in situ flow patterns immediately around the tool and in the surrounding deformed material. From their research, the authors concluded that there are five distinct material deformation regions associated with the formation of four-sided channels, which result from the dynamic interaction of the material with the FSC tool probe and shoulder.

Mehta and Vilaça reviewed variants of FSC that have emerged since Mishra's original disclosure [13]. Their review included related variants like the stationary shoulder approach recently advanced by Di Pietro et al. [14, 15]. FSC methods discussed included those in which tool tilt is set to zero and the shoulder face is maintained on the part surface. The reviewers found that FSC channels with a range in sizes could be fabricated to have acceptable internal surface finishes that qualify as being production-ready for certain applications. By combining friction stir welding (FSW) with FSC, a hybrid FSC process was shown to be capable of joining multiple parts together while forming channels along the joint line between both similar and dissimilar metal components [16, 17].

In early 2021, Burford and Mishra introduced a unique friction stir processing (FSP) technique for forming integral internal channels that utilizes specially designed bobbin tools [18]. This new approach for producing subsurface channels, called submerged bobbin tool (SBT) tunneling, involves positioning the shoulder at the distal end of a tool probe within the workpiece during processing. Conventional bobbin tools used for FSW, such as described by Goetze et al. [19], were not directly adaptable for this new patent pending use of friction stir bobbin tools. Therefore, specially designed bobbin tools were developed.

The initial SBT development was carried out on a CNC lathe equipped with a specially designed fixture as described by Burford et al. [20, 21]. This approach for forming channels was pursued because of the potential bobbin tools have for reducing out-of-plane process forces compared to single-sided friction stir tools. It was anticipated that this new approach would be capable of producing curvilinear internal pathways for a range of existing and new uses, including pathways for wiring, gases, fluids, and tubing, as well as reservoirs for powders and solid materials such as composites. Targeted components include heat exchangers, cooling plates, and vacuum tools. Other potential uses are in electric vehicle battery trays and in structural components having integral pathways for embedded sensor wiring, pneumatic actuator lines, and engineered designs for weight reduction and stiffness tuning.

Following the initial SBT development phase, a program for testing the transferability of SBT tunneling was carried out. For this next phase, several series of coupon-level tests were completed on two pieces of equipment located on the Discovery Park campus of the University of North Texas (UNT): (1) a purpose-built FSP machine and (2) an FSP-capable industrial robot. Results from the transferability phase and the prior developmental phase demonstrated that this new technology is adaptable to a range of industry equipment. SBT tunneling is therefore expected to potentially produce a variety of channel configurations for a range of applications—from linear channels to curvilinear internal pathways in complex-shaped parts.

A general description of the SBT tunneling process is presented in the next section. Results from the initial process development phase and the technology transfer assessment phase are then discussed and summarized in the subsequent sections.

SBT Tunneling

SBT tunneling is based on a unique tool design that has a probe with the general shape of a bobbin. The components of an example two-piece toolset are shown in Fig. 1. The shoulder at the terminal end of the probe is specially designed to pass efficiently through the workpiece as the process progresses. An important function of the probe design is to consolidate the FSP bridging material formed behind the tool as it advances along the path of the channel. This process is illustrated schematically in Fig. 2, which shows a cut-out around an advancing SBT tool.

Toolsets for SBT tunneling are designed to form integral subsurface channels (tunnels) having a machined-like surface finish on all internal faces, including on the bridge or ceiling face of a given channel. Like a conventional bobbin tool (BT) used in FSW, an SBT toolset has two opposing shoulders spaced apart along the bobbin or probe section of the tool. Unlike a conventional BT, an SBT is used to form enclosed internal channels by the distal shoulder—the shoulder located at the terminal end of the bobbin—being submerged within the workpiece while the opposite shoulder is positioned on an outer surface of the workpiece during processing [18, 21]. The

Fig. 1 Example set of preproduction SBT toolsets. Each tool is comprised of (1) the SBT probe and (2) the tool body with integral shoulder [16]

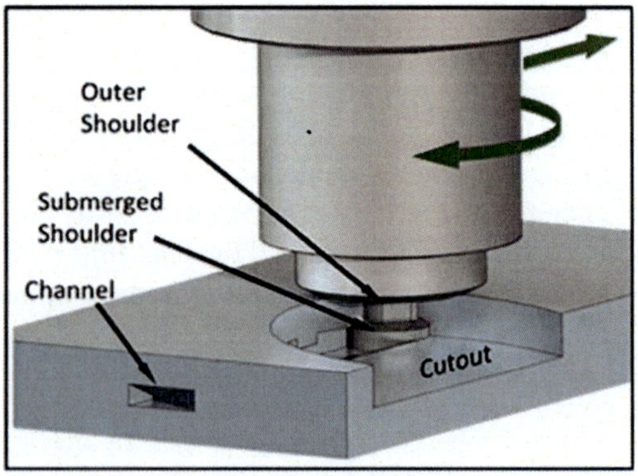

Fig. 2 Schematic of a cut-out around an advancing SBT tool for producing integral channels

shape of this unique form of tool design introduces physical support to the processed material as it recombines along the trailing region behind the rotating tool.

Similar to other BT designs, the opposing shoulders of SBT toolset designs serve to contain a significant portion of stirred material generated throughout the progression of the process. As a result, process forces generated parallel to the tool's axis of rotation are reacted between the opposing shoulders. Compared to single-sided tool designs having one shoulder, SBT toolsets therefore produce relatively lower out-of-plane forces that must be supported by the fabrication equipment in use. This, in turn, means that the fabrication equipment for the SBT process has reduced force and stiffness requirements compared to equipment for single-sided FSC methods. Also demonstrated in this work, SBT tunneling may be performed rapidly with an appropriate toolset design. For example, well-formed channels have consistently been produced in AA6061 aluminum at travel speeds of 635 mm/min, and speeds as high as 890 mm/min were achieved.

SBT tunneling was also designed to be a versatile process. Because of the narrow profile of SBT toolset designs, this unique process can be applied to complex part configurations. One example is illustrated schematically in Fig. 3, where a channel is "installed" in a confined space next to a flange. Additionally, sequencing SBT tunneling operations with machining operations on machining centers opens opportunities for completing parts in a single machine setup. Note, however, that SBT tunneling is not a machining operation. It is a localized thermomechanical process which depends upon process parameters in a much different way when compared to machining operations.

Figure 3 also illustrates how SBT channels may be used with or without inserts or liners. In either case, SBT channels can serve as conduits for wiring and for tubing which is used to distribute various media. For many applications, e.g., in pneumatic and/or vacuum lines, inserts and liners are not needed. Included in the illustration presented in Fig. 3 is a channel that serves as a conduit for routing wiring to a sensor or device, such as a proximity sensor. Other applications may include embedded sensors for measuring accelerations, vibrations, and impact. They may also be installed for measuring the content of gases, such as humidity or vapors from volatile substances.

Another example use of SBT channels is illustrated in Fig. 4. The component shown in the figure is an idealized modular out-of-autoclave vacuum tool having internal pathways for heating oils. Integral subsurface channels are also incorporated

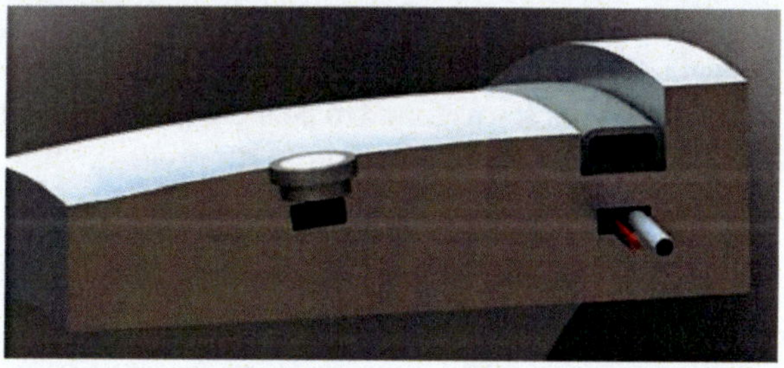

Fig. 3 A schematic of a generalized curvilinear part demonstrating optional combinations of integral channels, including placement of smart sensors, sensor wiring, tubing, etc

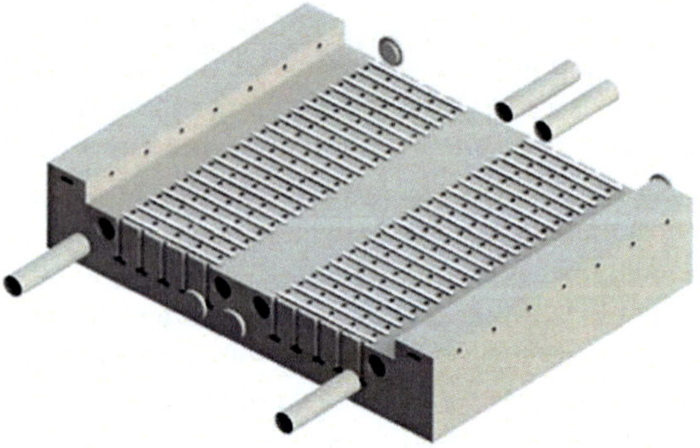

Fig. 4 An idealized modular out-of-autoclave vacuum tool having internal pathways for heating and cooling fluids. It incorporates other uses of integral subsurface channels, including vacuum lines and channels for thermocouple wire management

in the idealized design to include vacuum lines and channels for thermocouple wire management.

Similar uses of channels as those shown in this section may be designed into electric vehicle (EV) battery trays, structures for dense electronic components, and components designed for reduced heat signatures. Channels can also enable the functionality of part actuation and monitoring by providing integral pneumatic lines and pathways for distributing wiring to health monitoring devices, respectively. Structures having integral composite stiffeners are further examples of advanced uses of SBT integral channels. Parallel opportunities include the reduction of weight in monolithic structures in which channels cannot be extruded by conventional processes. As an FSP process, SBT may also be used to insert channels encased in wrought material into castings and parts manufactured via additive manufacturing. Repair of existing channels (e.g., in castings) as a remediatory process is another potential application of the SBT tunneling technology.

Phase 1: Initial Process Development

SBT tunneling was originally pursued to develop an agile, low-force approach to fabricating curvilinear integral pathways in a single pass. To meet this goal, a development program was carried out on a 2-D prototyping system pictured in Fig. 5. The main components of this 2-D system included: (1) a standard CNC lathe; (2) a set of experimental SBT toolsets (to be mounted in the chuck of the lathe); and (3) a custom fixture for holding FSP test coupons. Because of the limited force capabilities of this system, it was found to be ideal for driving the development of SBT tunneling toward being a low-force friction stir process.

Several other items are worth noting in Fig. 5. First are the 102 mm (4 in) long AA6061-T6511 coupons which are mounted in the FSP fixture: during processing (left) and after processing (right). The picture on the right also provides an example of the cone flash that is formed upon extracting the SBT probe at the end of the processing path. This flash is easily removed in subsequent machining operations.

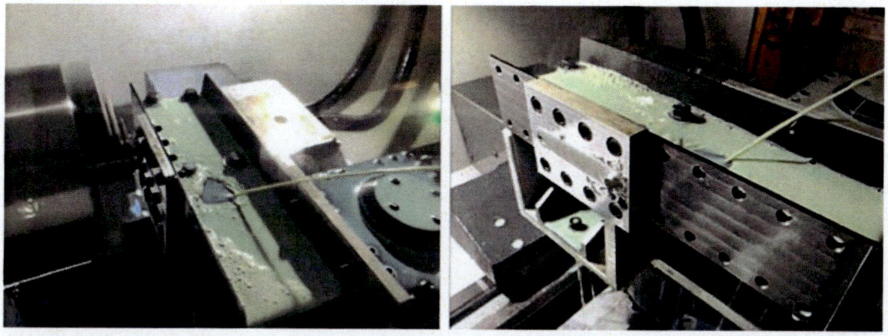

Fig. 5 2-D prototyping system comprised of a CNC lathe, experimental SBT toolsets (mounted in the lathe chuck), and test coupons mounted on a custom fixture built for friction stir tool development

Second, coolant is shown being applied to the backing anvil to dampen thermal transients in the fixture and thus reduce thermal expansion and contraction swings in its components during processing and the following cool-down phase. This is considered important for protecting the lathe equipment as well as for maintaining the integrity of the SBT process, which is run in position control on this test fixture.

Figure 6 provides details of a typical test coupon designed for the initial phase of developing preproduction SBT toolsets and related process parameters. The coupons were made from AA6061-T6511 extruded plate and ranged in thickness from 9.5 mm (0.375 in) to 25.4 mm (1.0 in). Key features of the coupon layout included a starter hole followed by a slot extending along the processing path. The slot was not directly connected to the starter hole to ensure there was sufficient area around the hole for initial shoulder contact, which is needed for uniform frictional heating at the start of the processing phase. This spacing was also sized to ensure sufficient material was available to fill the bobbin early in a process run. These machined features were adjusted in size and location as part of the development phase to manage material movement during processing.

Given the bobbin shape of an SBT probe, a starter hole was found to be necessary for beginning the tunneling process within the profile of the workpiece. The purpose of the slot was to provide a reservoir of sufficient capacity to receive the material that is displaced by the shoulder on the distal end of the probe. It was found that if the slot is too large, not enough material is available to form a fully consolidated bridge or ceiling over the channel. If the slot is too small, there is a risk of breaking the probe because of the buildup of process forces with the accumulation of material ahead of the traversing tool.

Initially, the two-piece SBT toolsets were sized for a nominal channel cross-section of 2 mm by 9 mm. To arrive at a stable channel cross-section along the entire length of the channel required a series of development iterations involving adjustments to the shoulder diameter, shoulder scrolls, and the features and shape of the SBT probe. Obtaining a well-formed channel from the start of a processing pass required careful control of the plunge and the tool dwell stages. Following the dwell

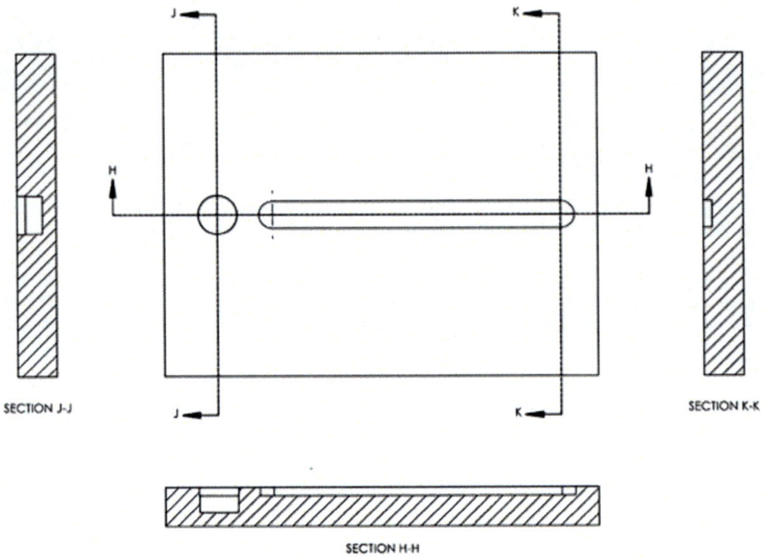

Fig. 6 Schematic of a test coupon for developing process parameters in AA6061-T6511 extruded plate

stage, a carefully controlled ramp-up to the steady-state travel speed was developed as shown in Fig. 7. In all cases, the transition to the steady-state travel speed occurred in less than 25 mm.

As noted previously, using a CNC lathe imposed certain constraints on the development phase. Aside from the 2-D motion limitation, it was limited to position control

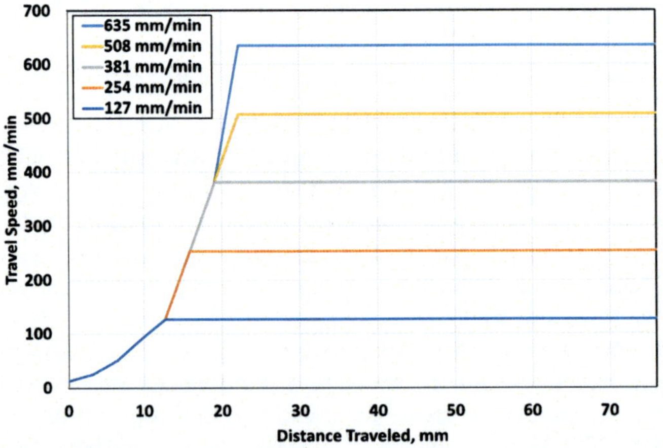

Fig. 7 A graph depicting the process sequence in terms of travel speed for five different travel speeds, ranging from 127 mm/min to 635 mm/min at 750 rpm

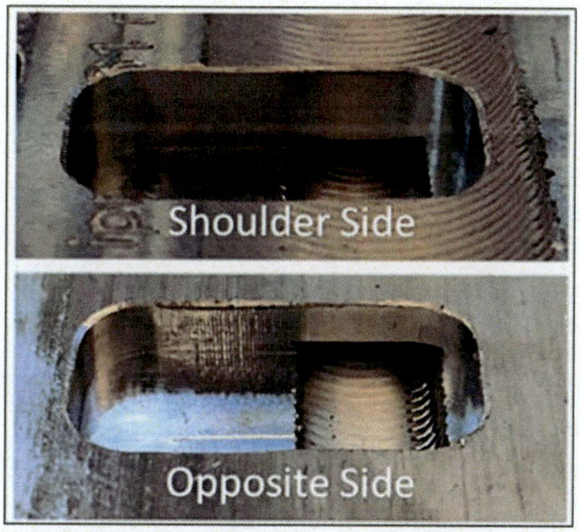

Fig. 8 Shoulder (top) side and opposite (bottom) side of a 6061-T6511 coupon processed at 610 mm/min (24 ipm) and 750 rpm and then machined to view the interior of the channels

and each of its axes was limited in power. These constraints governed the size of the channels and the travel speeds that could be attempted. During this initial phase, attempted processing speeds ranged from 25.4 mm/min to 635 mm/min. A series of probe styles were tested until a robust tool design was demonstrated to produce multiple successive channels at a travel speed of 635 mm/min.

Figure 8 shows slots which were machined in either side of a typical development coupon to visually inspect the walls of channels in processed coupons. Both the upper and the lower surfaces of the exposed channels were found to have the appearance of a machined surface, having semicircular striations similar to the shoulder tracks on the top surfaces of the coupons.

As part of documenting the quality of SBT channels, test coupons were examined visually, metallographically, and with an X-ray microscope (XRM). The XRM proved quite valuable for documenting the nature of the SBT channel surfaces, providing 3-D images like the one in Fig. 9 for characterizing surface roughness. As can be seen, the surface is regular or periodic. Slower travel speeds were found to produce closer spacings in the surface roughness periodicity.

Initial steps were taken to investigate the benefit of treating channel surfaces by abrasive flow machining (AFM). Figure 10 shows images and cross-sections of selected coupons treated with this internal honing process. FSP processing parameters for the channels shown in Fig. 10 were 254 mm/min, 381 mm/min, and 508 mm/min, respectively as indicated in the figure. In all cases, the spindle rotation rate was 750 rpm.

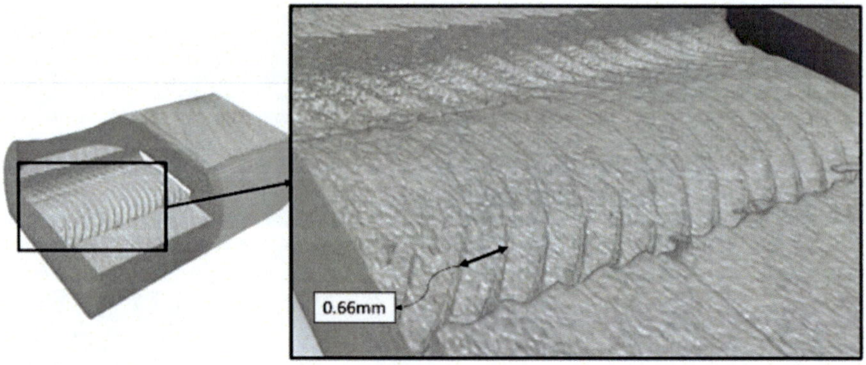

Fig. 9 Surface model built from XRM test data results for a channel produced at 635 mm/min (25 ipm) and 750 rpm

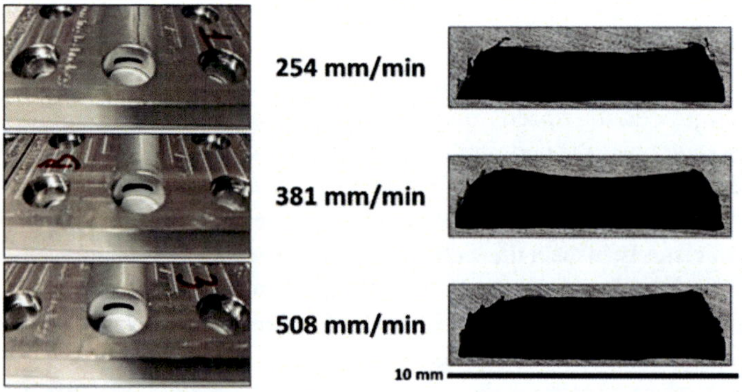

Fig. 10 Images and cross-sections of selected coupons treated with abrasive flow machining (AFM). FSP processing parameters for channel: 254 mm/min upper row; 381 mm/min middle row; and 508 mm/min bottom row. Spindle rotation speed was 750 rpm. For the AFM process, 600 psi was used with a 104 mm diameter machine cylinder. The abrasive flow media had a medium viscosity with an abrasive grit size of 150. Samples were processed between 4 to 12 cycles, with an average flow time per cycle of 6 min

Phase 2: Technology Transfer Evaluation

Two technology transfer paths for transferring SBT tunneling from the CNC lathe testbed are depicted in Fig. 11. Each was tested in the summer of 2021. The first was to a purpose-built FSP machine. The second was to an industrial robot equipped with an FSP end effector. Results of these evaluations are presented in the next two subsections.

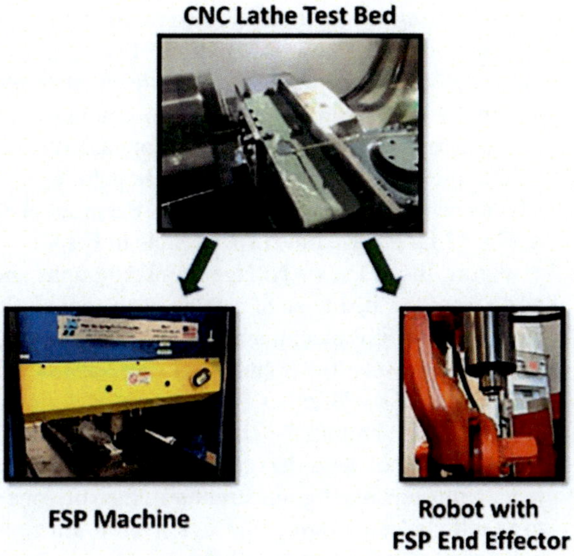

Fig. 11 Schematic showing the two technology transfer paths evaluated in this study: from the CNC lathe testbed to (1) to a purpose-built FSP machine and (2) an industrial robot equipped with an FSP end effector

Purpose-Built FSP Machine

A set of five coupons, shown in Fig. 12, were processed under position control on the UNT purpose-built FSP machine for the first technology transfer path test. This set of coupons was premachined to the same design and from the same stock material used in the development phase. Each test coupon was then processed at a travel speed corresponding to a speed used in Phase 1, ranging from 127 mm/min to 635 mm/min.

Figure 13 presents process forces plotted as a function of distance traveled and rate of travel for coupons processed at 254 mm/min, 381 mm/min, 508 mm/min, and 635 mm/min, respectively. The forge force for the steady-state portion of the process ranged from nominally 3.5 kN at 254 mm/min to nominally 4.7 kN at 635 mm/min.

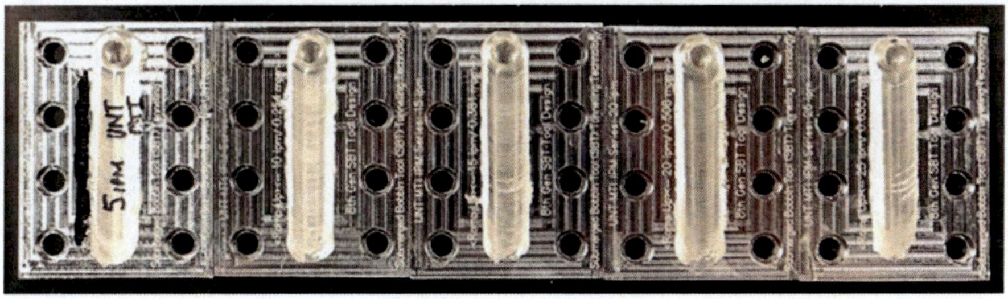

Fig. 12 Set of five SBT tunneling coupons processed on the UNT purpose-built FSP machine. Five individual travel speeds were tested: left to right, 127 mm/min (5 ipm), 254 mm/min (10 ipm), 381 mm/min (15 ipm), 508 mm/min (20 ipm), and 635 mm/min (25 ipm). The rotational rate for each coupon was 750 rpm

The traversing or longitudinal force for the steady-state portion of the process ranged from nominally 1.7 kN at 254 mm/min to nominally 2 kN at 635 mm/min. In addition, the cross or transverse force for the steady-state portion of the process ranged from nominally 0.7 kN at 254 mm/min to nominally 2 kN at 635 mm/min. The initial transient segment of the curves corresponds to the transition in travel from the starting hole to the slot along the SBT process path. It also relates to the thermal softening which occurs in the ramp-up phase of the process cycle. As expected, the force plots in Fig. 13 reflect the travel speed plots in Fig. 7.

Figure 14 presents a collage of macrographs and micrographs of a cross-section taken from the steady-state portion of the coupon processed at 381 mm/min on the purpose-built FSP machine at UNT. No voids were evident in any of the stir zones examined between the bobbin tool shoulder track and the bridge or ceiling face of the channel. As with other FSP operations, a fine grain microstructure is produced in the stir zone around the channel. While it is observed that the stir zone extending between the tool shoulders has a classical appearance of a friction stir morphology (e.g., a fine equiaxed grain structure), the stir zone produced by the tool tip appears to be minimal, suggesting this region of the stir zone may have limited impact on the parent material beneath the channel.

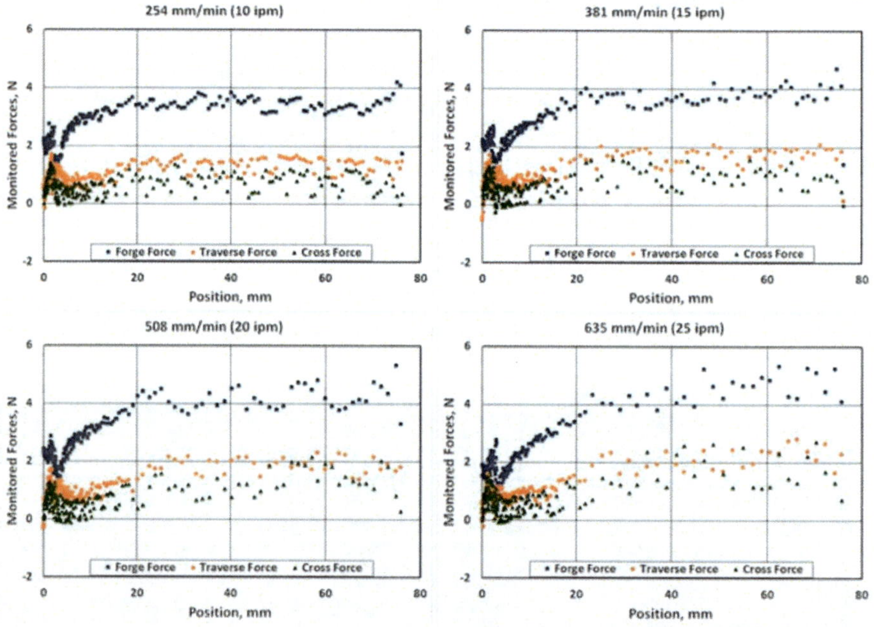

Fig. 13 Purpose-built FSP machine process forces versus distance traveled for four traversing speeds as noted: 254 mm/min, 381 mm/min, 508 mm/min, and 635 mm/min, respectively; all at a rotation rate of 750 rpm

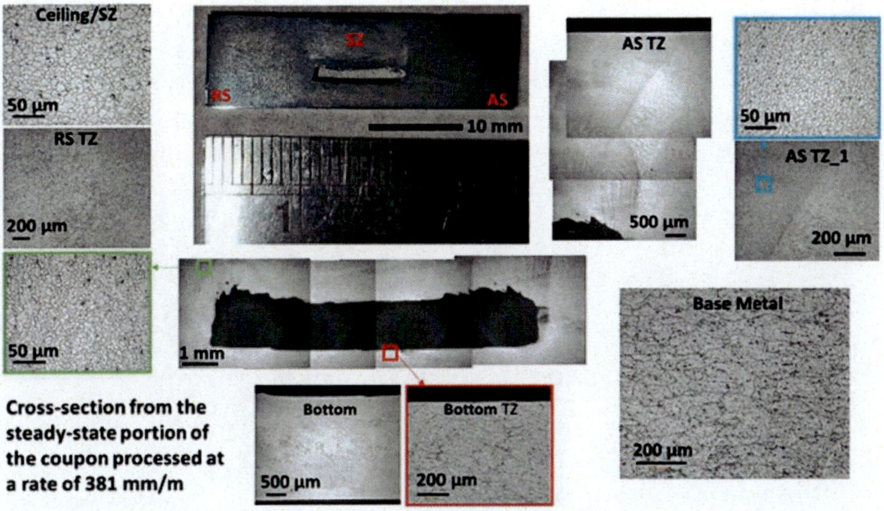

Fig. 14 Cross-sections from the steady-state portion of the coupon processed on the UNT purpose-built FSP system at a travel speed of 381 mm/min and rotational rate of 750 rpm

Robotic FSP System

Because the SBT tunneling process under investigation was found to produce relatively low process forces (Fig. 13), it was anticipated that it could be transferred directly to a properly rated industrial robotic system equipped with an FSP end effector. To test this supposition, a series of coupons identical to those tested on the FSP machine were processed with the industrial robotic system also located at UNT. A preproduction SBT toolset mounted on the robot arm is shown in Fig. 15.

The first set of SBT coupons processed on the robotic system is shown in Fig. 16. As expected, the tool deflected away from the programmed path at the start of the run, which was corrected manually as the run progressed. Based on the magnitude of these manual adjustments, the robot was then trained to make the adjustments automatically. A second attempt was made, which was successful in terms of the robotic arm following the programmed path as trained. This is demonstrated by comparing the pictures presented in Fig. 16.

For the first run, it was anticipated that the toolset could deflect in the vertical direction as well. Therefore, the robotic arm was programmed in position control to over-plunge the SBT shoulder into the part surface. After the first run, it was found that this step was not needed to ensure the shoulder was fully engaged with the surface. This is not an unexpected result since SBTs are internally self-reacting along the tool axis of rotation. In the next run, the magnitude of the over-plunge was lessened, and it was eventually reduced to zero for processing subsequent coupons. A major observation of these test runs was that even with an over-plunge, the SBT toolset

Fig. 15 An SBT toolset mounted on the UNT robotic FSP system

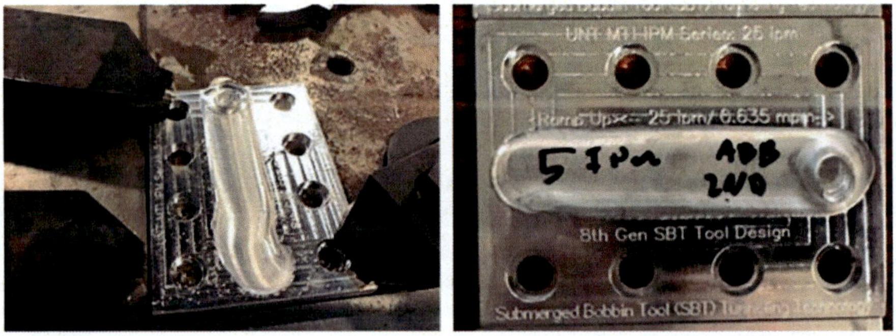

Fig. 16 First set of SBT test coupons processed on the UNT robotic FSP system: before positioning compensation (left) and after compensation (right). The travel speed was 127 mm/min, and the rotational rate was 750 rpm for both coupons

capably produced channels. This result signified that the SBT tunneling process is robust in that it can accommodate deviations from the programmed travel path.

Figure 17 shows SBT tunneling coupons processed with correction for side-deflection at 381 mm/min and 508 mm/min, respectively. A small measure of flash was observed in both coupons, which was easily removed by hand. The distortion in the cone-shaped flash at the exit holes gives a sense of the lateral deflection in the robotic arm resulting from in-plane process forces. The process path for each was again observed to be nominally straight.

The first attempt at processing a U-shaped channel is shown in Fig. 18 near the start of processing at 127 mm/min and 750 rpm. As observed, the figure shows the robotic arm initially deflected to the side of the path in response to in-plane process forces and some dynamic programming issues. By correcting the off-set displacement in

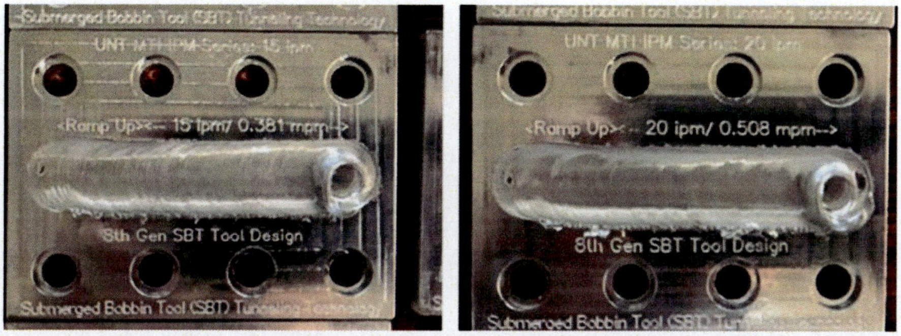

Fig. 17 SBT coupons processed with the UNT robotic FSP system at 381 mm/min (left) and 508 mm/min (right). Both coupons were processed at a spindle speed of 750 rpm

real-time, the robot was eventually able to follow the machined slot in the coupon. From prior processing runs on the purpose-built FSP machine, it was known that the side forces could have been on the order of 1 kN for a travel speed of 127 mm/min. For future runs, it is anticipated that correlating force data with deflection data will be useful in training the robot system for advanced development work on complex shapes in the robotic cell.

Figure 19 shows the second process run, which was also carried out at 127 mm/min and 750 rpm. This time the coupon was processed with side-deflection correction from the start of the path. As shown in the figure, the robotic arm was able to follow the scheduled processing path around the corner of the path to the end of the machined slot. Both channels produced at 127 mm/min were especially useful because they provided an opportunity to examine how channel configurations are affected by the robot arm straying from the anticipated processing path.

A third U-shaped channel was processed at 254 mm/min. This channel and the second U-shaped channel processed at 127 mm/min (with a correction factor) were sectioned as shown in Fig. 20 for metallographic examination—with the 127 mm/min coupon shown on the left and the 254 mm/min coupon shown on the right in this

Fig. 18 First attempt at processing U-shaped channels on the UNT robotic FSP system. This coupon was run at 750 rpm and 127 mm/min initially without side-deflection control

Fig. 19 Second U-shaped channel processed on the UNT robotic FSP system at 127 mm/min and 750 rpm. This process run included side-deflection correction

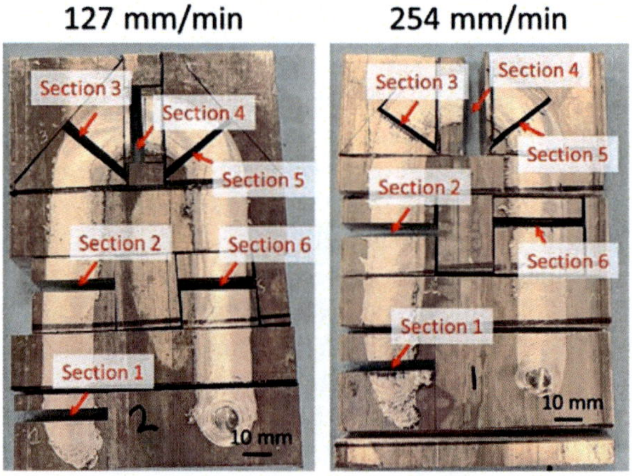

Fig. 20 Sectioning of the U-shaped channels processed on the UNT robotic FSP system at 127 mm/min (left) and 254 mm/min (right), respectively

figure. The cross-sections of sectioned samples taken from six locations in each coupon are shown in Fig. 21.

In Fig. 21, the channel morphology is shown to vary along the U-shaped path. In all positions examined along the path, a channel was produced, which serves to demonstrate the robustness of the technology. However, the morphology of the channel cross-section is observed to vary from one section to the next. Therefore, further research is needed to understand these preliminary results produced by the robotic system.

Discussion and Conclusions

SBT tunneling holds significant potential for producing channels in malleable materials for a host of applications because of its robustness and agility. Although the research covered in this paper has established the viability of the process, this research

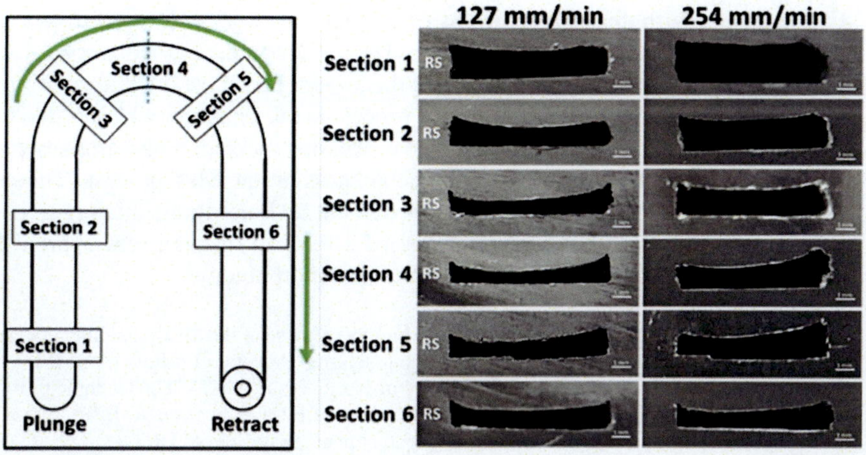

Fig. 21 Cross-sections of U-shaped channels processed at 127 mm/min (left) and 254 mm/min (right) on the robotic FSP system

was just the initial phase in developing this new technology. Additional alloys and channel shapes need testing, and SBT process parameters, toolsets, and fixturing need maturing for specific materials and products. A recent study of U-shaped channels produced by SBT tunneling highlights these needs [22].

From results obtained by Koonce [22], surface finishes in channels produced by SBT tunneling appear to be dependent upon the thermal condition and behavior of the workpiece under certain working conditions. For example, in reviewing the results obtained by Koonce [22], it was observed that thicker test plates appeared to have better channel surface finishes than those produced in thinner plates. This result points to a possible thermal effect since thicker plates can absorb and conduct a greater portion of heat generated by processing compared to thinner plates. It was also observed that channels produced by the purpose-built FSP machine were found to have worse surface finishes compared to those produced in the robotic cell.

This difference in surface finishes produced by the two systems does not appear to be explained by the processing equipment itself. Even though the purpose-built FSP machine was stiffer than the robot—and therefore the SBT process was better controlled by it—the robot produced better surface finishes within the channels using the same SBT toolset. To resolve the observed differences in surface quality, the fixture in both systems was examined. The purpose-built FSP machine (shown in Fig. 11) was originally equipped with an O1 tool steel backing plate that was not actively cooled for this work. A thick plate of uncooled A36 steel was used for a fixture in the robot cell (e.g., Fig. 19). For comparison, in the CNC lathe testbed (Fig. 5), coupons were supported by an actively cooled support beam made of A36 steel.

The influence of the backing plate was considered in follow-on work to the present study where a copper backing plate was placed over the O1 plate in the purpose-built

FSP machine. With this change, it was found that comparable surface finishes could then be obtained between the two systems. The result obtained by Koonce [22] and in the follow-up work to this study highlight the importance of the thermal environment immediately around where a friction stir process is taking place, which is due to the localized nature of this thermomechanical processing method. It also emphasizes the need to evaluate the effect of workpiece thickness on the SBT process. Therefore, further work like the research program carried out by Balasubramanian et al. [8–11] would serve to advance the development of a thermal management approach for more directly controlling channel formation and morphology.

Acknowledgements Funding for the initial development phase of the SBT tunneling process was provided through the Advanced Materials & Manufacturing Processes Institute (AMMPI) located at the Discovery Park campus of the University of North Texas (UNT). The technology transfer phase conducted during the summer of 2021 was funded by a grant from AMMPI. Submerged Bobbin Tool (SBT) Tunneling Technology is a patent pending process of UNT.

References

1. Mishra R (2004) Integral channels in metal components and fabrication thereof. US Patent Application Publication 20040060965A1. 1 April 2004
2. Mishra RS (2005) Integral channels in metal components and fabrication thereof. US Patent 6,923,362. 2 August 2005
3. Mishra RS (2008) Integral channels in metal components and fabrication thereof. US Patent 7,354,657. 8 April 2008
4. Ayub ZH (2003) Plate heat exchanger literature survey and new heat transfer and pressure drop correlations for refrigerant evaporators. Heat Transfer Eng 24(5):3–16. https://doi.org/10.1080/01457630304056
5. Miller WS, Zhuang L, Bottema J, Wittebrood AJ, De Smet P, Haszler A, Vieregge A (2000) Recent development in aluminium alloys for the automotive industry. Mater Sci Eng, A 280(1):37–49. https://doi.org/10.1016/S0921-5093(99)00653-X
6. Harris C (2001) Design, fabrication, and testing of cross flow micro heat exchangers. Doctor of Philosophy. Louisiana State University and Agricultural & Mechanical College. https://doi.org/10.31390/gradschool_disstheses.290
7. Launay S, Sartre V, Bonjour J (2007) Parametric analysis of loop heat pipe operation: a literature review. Int J Therm Sci 46(7):621–636. https://doi.org/10.1016/j.ijthermalsci.2006.11.007
8. Balasubramanian N, Krishnamurthy K, Mishra RS (22 May 2009) Preliminary study of pressure drop and heat transfer through a friction stir channel. ASME 2007 international mechanical engineering congress and exposition. Seattle, Washington, USA pp 933–939. https://doi.org/10.1115/IMECE2007-41634
9. Balasubramanian N, Mishra RS, Krishnamurthy K (2009) Friction stir channeling: Characterization of the channels. J Mater Process Technol 209(8):3696–3704. https://doi.org/10.1016/j.jmatprotec.2008.08.036
10. Balasubramanian N, Mishra RS, Krishnamurthy K (2010) Development of a mechanistic model for friction stir channeling. J Manuf Sci Eng 132(5):054504. https://doi.org/10.1115/1.4002453
11. Balasubramanian N, Mishra RS, Krishnamurthy K (2011) Process forces during friction stir channeling in an aluminum alloy. J Mater Process Technol 211(2):305–311. https://doi.org/10.1016/j.jmatprotec.2010.10.005

12. Pandya S, Mishra RS, Arora A (2019) Channel formation during friction stir channeling process—a material flow study using X-Ray micro-computed tomography and optical microscopy. J Manuf Process 41:48–55. https://doi.org/10.1016/j.jmapro.2019.03.021
13. Mehta KP, Vilaça P (2022) A review on friction stir-based channeling. Crit Rev Solid State Mater Sci 47(1):1–45. https://doi.org/10.1080/10408436.2021.1886042
14. Di Pietro V, Gandra J, Dodds S (3 May 2021) Thermal management of power components and electric systems using channels embedded in metallic parts by friction stir channelling. PCIM Europe digital days 2021; international exhibition and conference for power electronics, intelligent motion, renewable energy and energy management. VDE VERLAG GMBH, Berlin, Offenbach, pp 1138–1143
15. Di Pietro V, Gandra J, Clarke A, Hockaday J, Law R (March 2022) Removing design and manufacturing complexity from integrated thermal management solutions using friction stir channelling. presented at the 2022 38th semiconductor thermal measurement, modeling and management symposium (SEMI-THERM). San Jose, CA. https://ieeexplore.ieee.org/document/9755866
16. Karvinen H, Nordal D, da Silva PSV (2018) Microstructural characterization of the stir processed zones of Cu-Al and Al-Al plate systems manufactured by hybrid friction stir channeling. 12th International symposium on friction stir welding. TWI Ltd. Session 7 (Techniques II). 26 June 2018. https://research.aalto.fi/en/publications/microstructural-characterization-of-the-stir-processed-zones-of-c
17. Karvinen H, Nordal D, Galkin T, Vilaça P (2018) Application of hybrid friction stir channeling technique to improve the cooling efficiency of electronic components. Welding World 62(3):497–509. https://doi.org/10.1007/s40194-018-0576-8
18. Burford DA, Mishra R (28 July 2022) Systems and methods for internal channel formation within a workpiece. US Patent Application Publication 20220234132A1
19. Goetze P, Kopyściański M, Hamilton C, Dymek S (12 Feb 2019) Comparison of dissimilar aluminum alloys joined by friction stir welding with conventional and bobbin tools. In: Friction stir welding and processing X, the minerals, metals and materials series. Springer, Cham pp 3–12. https://doi.org/10.1007/978-3-030-05752-7_1
20. Burford D (15 March 2022) Submerged bobbin tool (SBT) tunneling technology. presented at the AeroMat 2022. Pasadena, CA. https://asm.confex.com/asm/aero22/webprogram/Paper55159.html
21. Burford D (2022) Submerged bobbin tool (SBT) tunneling technology. presented at the AeroMat 2022: research and development enabling next-generation aerospace materials. Pasadena, CA. https://www.youtube.com/watch_popup?v=49mS2jZyAN4
22. Koonce JG (2022) Impact of processing parameters and forces on channels created by friction stir bobbin tools, master's thesis. University of North Texas (UNT). https://digital.library.unt.edu/ark:/67531/metadc1944372/

Author Index

Y. Hovanski et al. (eds.), *Friction Stir Welding and Processing XII*,
The Minerals, Metals & Materials Series,
https://doi.org/10.1007/978-3-031-22661-8

Subject Index

Y. Hovanski et al. (eds.), *Friction Stir Welding and Processing XII*,
The Minerals, Metals & Materials Series,
https://doi.org/10.1007/978-3-031-22661-8